Science Museums in Transition

Science Museums in Transition: Unheard Voices considers how museums can adapt their exhibits, programs, and organizational structures to the diversity of ideas, people, and cultures that speak to modern science.

This collection contains individual expressions by museum insiders addressing a range of particular perspectives – Native American, African American, Latin American, Islamic, Israeli, Danish, and white North American. These reflections provide guidance to the museum community as to how their institutions can become more thoughtful, more welcoming to diverse audiences, and more cognizant of the ways that different people incorporate science into their daily lives. As a whole, the book emphasizes the need for museums to engage in dialogue with their visitors – not merely to present them with information – and to offer opportunities to share experiences and exchange perspectives, and thereby advance science learning through a dynamic and collective process.

Science Museums in Transition is intended to further discussion on how museums address the political and social ramifications of science and, as such, should be of great interest to academics, researchers, and postgraduate students working in the fields of museum studies, science, anthropology, education, and history. It should also be essential reading for museum professionals around the globe.

Hooley McLaughlin is an Adjunct Professor in the Faculty of Information Studies at the University of Toronto in Canada.

Judy Diamond is a Professor and Curator at the University of Nebraska State Museum in the USA.

Museums in Focus
Series Editor: Kylie Message
The Australian National University, Australia

Committed to the articulation of big, even risky, ideas in small format publications, 'Museums in Focus' challenges authors and readers to experiment with, innovate, and press museums and the intellectual frameworks through which we view these. It offers a platform for approaches that radically rethink the relationships between cultural and intellectual dissent and crisis and debates about museums, politics and the broader public sphere.

'Museums in Focus' is motivated by the intellectual hypothesis that museums are not innately 'useful', safe' or even 'public' places, and that recalibrating our thinking about them might benefit from adopting a more radical and oppositional form of logic and approach. Examining this problem requires a level of comfort with (or at least tolerance of) the idea of crisis, dissent, protest and radical thinking, and authors might benefit from considering how cultural and intellectual crisis, regeneration and anxiety have been dealt with in other disciplines and contexts.

Global Trends in Museum Diplomacy
Post-Guggenheim Developments
Natalia Grincheva

A Museum in Public
Re-visioning Canada's Royal Ontario Museum
Susan Ashley

Science Museums in Transition
Unheard Voices
Edited by Hooley McLaughlin and Judy Diamond

www.routledge.com/Museums-in-Focus/book-series/MIF

MUSEUMS IN FOCUS

Science Museums in Transition

Unheard Voices

**Edited by Hooley McLaughlin
and Judy Diamond**

LONDON AND NEW YORK

First published 2020
by Routledge
2 Park Square, Milton Park, Abingdon, Oxon OX14 4RN

and by Routledge
605 Third Avenue, New York, NY 10017

First issued in paperback 2021

Routledge is an imprint of the Taylor & Francis Group, an informa business

British Library Cataloguing-in-Publication Data
A catalogue record for this book is available from the British Library

Library of Congress Cataloging-in-Publication Data
A catalog record for this book has been requested

ISBN 13: 978-0-367-78775-2 (pbk)
ISBN 13: 978-1-138-48997-4 (hbk)

Typeset in Times New Roman
by Apex CoVantage, LLC

This book is dedicated to the youth of today
who will create the museums of tomorrow

Contents

Contributor biographies

Marianne Achiam is Associate Professor in the Department of Science Education at the University of Copenhagen, Denmark. She studies the cognitive implications of exhibit design and engineering and has authored the article "Informal Science Education and Gender Inclusion" (2017, L. S. Heuling ed., *Embracing the Other: How the Inclusive Classroom Brings Fresh Ideas to Science and Education*). She also co-edited *Educational Design in Math and Science: The Collective Aspect* (2016). She currently leads with Jeff Dodick (from the International Institute for Policy Research) the international project: Science//Museums: Science Museums, Youth and Diversity, funded by the International Network Programme of the Danish Agency for Science and Higher Education.

David Begay, Ph.D., is a principal in the Indigenous Education Institute and co-author of the book, *Sharing the Skies: Navajo Astronomy* (2006). With Nancy C. Maryboy, he has led a number of NSF-funded projects including the *Cosmic Serpent: Bridging Native and Western Learning in Museum Settings* project. They are also contributors to the Yale School of Forestry and Environmental Studies course, Journey Conversation: Weaving Knowledge and Action: Indigenous Ways of Knowing.

Elijah Benson (Mandan Hidatsa Arikara Nation) is a member of the MHA Nation Interpretive Center Steering Committee and a Nueta Language Apprentice for the MHA Nation Department of Education. He has been documenting MHA tribal members for the last five years in multiple mediums of interpretation.

Ron Blonder, Ph.D., is the head of the chemistry group in the Department of Science Teaching in the Weizmann Institute of Science. She is concerned about two central dimensions in chemistry education: connecting high school chemistry to contemporary research in chemistry and supporting teachers in integrating technology in chemistry teaching and addresses these concerns with research and envelopment. She led the Weizmann Institute's team in the Irresistible Project.

Rana Dajani is a molecular biologist, a Fulbright Fellow, Eisenhower fellow, Associate Professor at Hashemite University, Jordan, and a Yale and Cambridge visiting professor. A higher education reform expert, a member of the UN Women Jordan Advisory Council, she frequently writes in *Nature* about science, education and women in the Arab world. She established a network for women mentors which received the PEER award 2014. Other awards include: Most Influential Women Scientists in Islamic World; twelfth among the 100 most influential Arab women in 2015 and in 2017; and Global Change-maker Award from IIE/Fulbright. She has developed a community-based model "We love reading" to encourage children to read for pleasure which resulted in her receiving the Synergos award for Arab World Social Innovators in 2009, the Library of Congress Literary Award 2013 Best Practices, WISE Award, King Hussein Medal of Honor 2014, and Star Award for education impact in 2015, IDEO.org for best education program for refugees in 2015. WLR has spread to 30 countries.

Judy Diamond is Professor and Curator at the University of Nebraska State Museum. A fellow of the American Association for the Advancement of Science, she led the award winning *Wonderwise: Women in Science* project, the multi-institutional Explore Evolution exhibitions, and the *World of Viruses* outreach that resulted in the comic books, *Carnival of Contagion* and *World of Viruses*, and the Carl Zimmer book, *A Planet of Viruses*. She co-authored *Concealing Coloration in Animals* (2013) and *Thinking Like a Parrot: Perspectives from the Wild* (2019).

Nancy C. Maryboy, Ph.D., is a principal in the Indigenous Education Institute and co-author of the book, *Sharing the Skies: Navajo Astronomy* (2006). With David Begay, she has led a number of NSF-funded projects including the Cosmic Serpent: Bridging Native and Western Learning in Museum Settings Project. They are also contributors to the Yale School of Forestry and Environmental Studies course, Journey Conversation: Weaving Knowledge and Action: Indigenous Ways of Knowing.

Hooley McLaughlin is Adjunct Professor in the Faculty of Information at the University of Toronto. He recently retired from the Ontario Science Centre (OSC), Toronto, where he held the joint positions of Vice President for Science Experience and Chief Science Officer. He was at the OSC for thirty years, leading over a dozen major international exhibitions, including A Question of Truth, an investigation into cross-cultural scientific concepts and the history of bias and atrocities associated with the misuse of science.

Laura Huerta Migus is the Executive Director of the Association of Children's Museums. For more than fifteen years, she has spearheaded efforts to make museum experiences more equitable, with a particular emphasis on incorporating marginalized voices into museum narratives. Previous appointments include the Association of Science-Technology Centers, Inc., the National Multicultural Institute, and the National Association for Bilingual Education.

Mary Baker Price (Mandan Hidatsa Arikara Nation) is a Hidatsa Apprentice in the Culture and Language Division of the MHA Nation Department of Education. Mary recently received her BA in Native American Studies at the Nueta Hidatsa Sahnish College. She is also an AmeriCorp Alumni who was stationed at the Knife River Indian Villages National Historic Site in Staton, North Dakota as a Tribal VISTA in collaboration with the Southwest Conservation Corps Ancestral Lands Program, the program is meant to reconnect native youth to their cultural heritage and ancestral lands. Her focus is language and culture revitalization for the MHA Nation under the Culture and Language Division.

Sherman Rosenfeld works in the Department of Science Teaching in the Weizmann Institute of Science. Over three decades, Sherman has developed curricula, teacher education programs, and science education research. His career includes a science museum directorship in California and publications on bridging the gap between formal and informal science learning. He directed the Israeli Ministry of Education's Department of Science-Oriented Youth and participated in the EU-funded Irresistible Project.

Monique Scott, Director of Museums Studies at Bryn Mawr, is author of *Rethinking Evolution in the Museum*. She received her Ph.D. in anthropology from Yale and worked for a decade as head of cultural education at the American Museum of Natural History. Her chapter discusses how exhibits can transparently address colonialism from an interview with Tukufu Zuberi, Lasry Family Professor of Race Relations, and Professor of Sociology and Africana Studies at the University of Pennsylvania.

Jen Shannon is Curator and Associate Professor of Cultural Anthropology at the University of Colorado, specializing in collaborative anthropology and connecting Native communities to museums through repatriation, digitization of collections, and co-directed research projects. She is a contributor to *Sapiens: A Podcast for Everything Human*, and *Collaborative Exhibition Making: Thinking About Different Models for Community Participation* (2016). She is also co-author of the 2017 graphic narrative about the Native American Graves Protection and Repatriation Act of 1990, *Journeys to Complete the Work: Stories about Repatriations and Changing the Way We Bring Native American Ancestors Home.*

Royce Young Wolf (Freeman) (Hidatsa, Mandan, and Eastern Shoshone) is a member of the MHA Nation Interpretive Center Steering Committee and Culture & Language Division Advisor in the MHA Nation Tribal Education Department. She is an avid Hidatsa language student and is a Sociocultural and Linguistic Doctoral Candidate at the University of Oklahoma.

Tukufu Zuberi is Lasry Family Professor of Race Relations, and Professor of Sociology and Africana Studies at the University of Pennsylvania. Dr. Zuberi is a renowned documentarian on Africa and the African diaspora, and curator of exhibitions that challenge us to enhance our understanding of the African and African American experience. He is dedicated to bringing a fresh view of culture and society to the public through various platforms such as guest lecturing at universities, television programs, and interactive social media. Currently, he works on human rights initiatives by participating in public speaking engagements, international collaborations with transnational organizations, and individuals dedicated to human equality. Dr. Zuberi's research focuses on race, demography, and culture among African and African diaspora populations. In 2002, he became the founding Director of the Center for Africana Studies.

1 Inclusion and relevance in natural history museums

Judy Diamond

Enter the backrooms of any mid-to-large-sized natural history museum and view a magical world of singular specimens and unusual artefacts. Long rows of metal cabinets contain objects that range from the Earth's rarest to the most common and familiar. The contents of the cabinets and shelves constitute the intellectual foundation of the museum – items classified by taxonomic relationship, age, and sometimes geography, nation, culture, or tribe. There are mounts in drawers, fossils on shelves, wet specimens submerged in formalin, dried plants affixed to acid-free herbarium sheets, and artefacts cradled in foam stands. The collection, naming, and preservation of specimens and the artefacts of material culture delimit the original primary function of natural history museums. Specimens contain slices of the world's biological diversity in a banked reservoir of DNA. Cultural artefacts embody the ideas and practices of diverse peoples, recording and retracing the evolution of material culture. These collections are resources that allow students and scientists to discover the patterns of life history – identification, diversity, evolution, and sustainability. Life-forms that are not named are not known with certainty to exist. Cultures are invisible or forgotten without the description of their artefacts. It is the mission of research in natural history museums not only to order life's diversity, but also to make possible its discovery.

This inner world of the natural history museum is a private place, reserved for the community of researchers who contribute to the accumulation of knowledge of life's diversity and material culture. These collections are shared across the globe in a collective commitment to advancing scientific understanding. The care and preservation of the collections remains as much the function of natural history museums as it was in 1683 when the Ashmolean Museum in England became the first of its kind to open its doors to the public. At first glance, these collections appear timeless, enshrined within the protected walls of their museums. But the foundation of natural history museum collections is no longer stable. As funding from state and federal sources plummets, there is a new sense of siege, and in the present-day push and shove of competing interests, the long-term preservation of museum collections ranks low on the horizon of most politicians.

Natural history museums also have another mission that is reflected in an outer core of priorities and activities. This outer museum is nothing like the solemn, research-oriented sanctuaries of the collections. It is busy, loud, boisterous, artistic, and often commercial. The museum markets itself through its exhibits, programs, public relations, and fund-raising. This outer museum has multiple aims: financial sustainability, community support, education of school children, outreach, and visitation by the public.

Education in some form, or at least the dissemination of ideas, is part of the mission of most natural history museums. Some promote appreciation for science or teach scientific principles. One of the world's largest natural history museums, the American Museum of Natural History, has the mission: "To discover, interpret, and disseminate – through scientific research and education – knowledge about human cultures, the natural world, and the universe." The Smithsonian's National Museum of Natural History does not directly mention education in its mission statement: "Understanding the natural world and our place in it." But under the original mandate of James Smithson, the mission called for, "an establishment for the increase and diffusion of knowledge among men [*sic*]."

Most natural history museums take a traditional approach to education, using techniques that have not changed since the mid-nineteenth century. As such, education in museums can often feel like a one-way dissemination of ideas: knowledge is dispensed from the museum to students and the public. Outside the realm of museums, American educational systems have vacillated between traditional and progressive forms of education. In traditional education, the teacher or other authority figure dispenses knowledge, usually in the form of factual information. But for more than a century, influenced by John Dewey and others, many educational systems have instead favoured progressive approaches where teachers provide education resources for students to engage in their own intellectual discoveries, usually through inquiry-based activities or social learning. Students are encouraged to create personal meanings, learn in different styles, and think critically and independently.

Natural history museums have generally resisted progressive approaches in their education initiatives. Exhibits reflect the collections, focusing on specimens and artefacts, and labels present factual information about the objects displayed. In this sense, the exhibits effectively give students and the public a sense of the inner museum and thus attempt to make a broader audience aware of collections and research. There is no question that remarkable objects can have powerful emotional impacts. There are recollections from famous scientists that describe how a visit to a natural history museum inspired them to pursue a career in research. The National Museum of Natural History recently displayed an exhibit called, *Objects of Wonder*, featuring some of the most unusual and beautiful items from their collections. The goal was to generate awe and to create recognition for the uniqueness and the importance of the museum's collections. Like society's other iconic objects, such as fine art and architecture, items from the Smithsonian's collections clearly deserve appreciation.

Large natural history museums have put their awe-inspiring objects to good use, and they are among the most visited among all types of public institutions. But they tend to stumble when it comes to educating students and the public about scientific processes and the nature of scientific inquiry. Museum collections serve as essential reservoirs for the advancement of scientific understanding – through studies of evolution, biodiversity, and cultural diversity. But there are relatively few museum exhibits that effectively educate audiences about these ideas. Natural history museum curators often communicate the scientific aspects of their collections as definitive and static knowledge – an assemblage of facts created by mostly white, male scientists. The life sciences are presented as a set of known quantities defined by the collections, with few clues to how species form, how their populations are maintained, and how they respond to environmental change.

Outside the museum, there is a broad consensus among scientists and educators for a different view: science is not an assembly of facts, but a process of inquiry, persistence, and discovery. Modern science is fundamentally tentative and cooperative, methodologically diverse and embedded into the larger issues of the surrounding culture.[1] To encourage scientific understanding is to create connections between phenomena and personal experience. Scientific understanding involves discussion, argument, reflection, and synthesis – challenging norms and reflecting on bias.

Science is today's currency for empowerment and social change, and science and technology represent powerful forces that shape the lives of every human on the planet. Yet natural history museums do little to empower the public by making them well informed and infused with a critical understanding of the strengths and limitations of the processes underlying scientific information. The presentation of specimens and artefacts with labels providing identification and facts discourages scientific reasoning, limiting the available avenues for visitors to make personal connections. The method of presentation implies that naming species is the goal of scientific investigation rather than a component of a more dynamic process of inquiry.

Powerful collaborative approaches to exhibit design, where scientists, artists, and educators come together to build on each other's ideas have been advocated for decades by institutions like the Kellogg Foundation and by visionaries like Frank Oppenheimer and Michael Spock.[2] Modelled over the years, natural history museums regularly adopt "team" approaches to exhibit development, but in practice these groups are often crippled by imbalances of power and influence. Team members defer to curators' notions of what information to cover, ceding control of content as if it were a commodity to be accumulated and doled out. Too few museum educators and exhibit designers have the clout or resources to effectively develop the kinds of learning tools that encourage users to reason in new ways.

Some natural history museums are trying to modernize and reformulate their definitions of education to be more in tune with contemporary practices, and their new aims include terms like learner-focused, promoting discovery, empowering learners, and encouraging critical thinkers. But the transition from disseminating knowledge to empowering learners requires more than a change in aims; it is a radical shift, and it requires institutional transformation which, for the most part, has not occurred. Educators and designers contribute experiential knowledge but often with insufficient opportunity to integrate social and cultural awareness and the cognitive processes that drive and expand thinking and learning. And sadly, researchers continue to document how the implicit influences of social class and ethnic exclusion hinder the museum science experiences of students living in poverty and those from minority or immigrant backgrounds[3].

Embracing the processes of science requires a change in how museums understand their role as science communicators, embracing cultural and cognitively focused teaching – where the institutions build exhibits and programs based on an understanding of how diverse visitors reason about phenomena, not just on what they *should* know. What would natural history museum exhibits look like if their educational role was to create independently minded science learners? Would exhibits not only stimulate curiosity but also teach visitors to reflect, synthesize, advocate, challenge, and seek connections between social processes and natural phenomena? It is not a simple matter for a museum to engage visitors with science as it is practiced in real-world settings. Creating effective displays requires bringing visitors into the realm of scientific thinking through inquiry, insight, identity formation, and connections to existing beliefs and experience. The challenge is not only how to display and teach with objects but also how to make them relevant, and in this way, make it possible for visitors to scaffold new experiences onto what is already familiar and valued. Museum staff also need to recognize and respect the diverse and often unfamiliar ways that visitors make meaning from museum objects. Relevance is not just a political or social construct – it reflects a cognitive process that emerges in supportive environments that make learning possible – and it serves as a gateway to lifelong learning in school, the community, and the natural world.

The essays in this volume illustrate how science museums can progress toward inclusivity. The term "science museum" incorporates all of the various kinds of institutions that consider public science education in their mission. In the broadest sense, this includes not only natural history museums, but also science centres, nature centres, botanical gardens, aquaria, and zoos. The essays elucidate the barriers present in these institutions that have resulted in social exclusion and nonparticipation by diverse audiences. They also describe pathways for improvement, presenting exemplary programs that can serve as guides for all museums. The presence of enduring partnerships with local and diverse communities leads to a recognition that there is more than one way to learn science. When Western historical concepts are presented side-by-side with Indigenous and traditional understanding, museums enable all people to see the relevancy of authentic processes of science. This doesn't imply "doing for," since inclusion demands "doing with," and it includes being willing to change practices in order to accommodate different perspectives. But partnerships do not replace the need for museums to build their core leadership and staffing to reflect their local community and a diversity of cultures.

As museums continue their mission of preserving and displaying collections, the essays in this book show how their broader impacts can be strengthened to incorporate relevance, advocacy, and social responsibility. The concept of social justice is highly applicable to science museums as they consider how they reinforce the distribution of privilege and access with their communities. This collection doesn't attempt to be comprehensive of the voices that need to be heard in museums. But the richness of the conversations conveys a vision of how science museums can achieve long-term sustainability, not through individual advancement but by embracing diversity more deeply and more centrally in all aspects of their operations.

Notes

1 Tsybulsky, Dina, Dodick, Jeff, and Camhi, Jeff (2017). The effect of field trips to university research labs on Israeli high school students NOS understanding. *Research in Science Education*, 1–26. Dagher, Zoubeida R., and Erduran, Sibel (2016). Reconceptualizing the nature of science for science education. *Science & Education*, 25(1–2): 147–164. Abd-El-Khalick, Fouad (2012). Examining the sources for our understandings about science: Enduring conflations and critical issues in research on nature of science in science education. *International Journal of Science Education*, 34(3): 353–374.
2 Oppenheimer, Frank (1972). The exploratorium: A playful museum combines perception and art in science education. *American Journal of Physics*, 40: 978–984. Spock, Michael (1988). What's going on here: Exploring some of the more

elusive, subtle signs of science learning. In *Science Learning in the Informal Setting: Symposium Proceedings*, edited by Paul G. Heltne and Linda A. Marquardt, 254–261. Chicago: The Chicago Academy of Sciences.

3 Dawson, Emily (2004). "Not designed for us": How science museums and science centers socially exclude low-income minority ethnic groups. *Science Education*, 98(6): 981–1008. Brown, Bryan A. (2004). Discursive identity: Assimilation into the culture of science and its implications for minority students. *Journal of Research in Science Teaching*, 41(8): 810–834.

2 Re-exhibiting Africa at the Penn Museum

An interview with Penn Museum curator Tukufu Zuberi

Monique Scott

Museums have historically reduced Africa and African objects to a seemingly static, dark, and dusty prehistory. African peoples, subsequently, are often denied the modernism, complexity, and dynamism they so richly deserve.[1] I've studied representations of Africa in anthropology exhibitions for almost 20 years and worked in anthropological education at the American Museum of Natural History for almost ten years. For much of that time, I was frustrated by the lack of opportunities to create new exhibitions that represent Africa. It was incredibly rewarding, then, to become a curatorial advisor to the Penn Museum's renovation of their permanent African gallery, an inspiring and daunting undertaking. The new exhibition, scheduled to open on 28 September 2019, will showcase the Museum's rich African collection, which is among the largest African ethnographic collections in the country.[2]

The Curator of the new Penn Museum's African gallery is the distinguished Sociologist, Tukufu Zuberi, the Lasry Family Professor of Race Relations and Professor of Sociology and Africana Studies at Penn. From the first moment Dr. Zuberi recruited me to the project, he spoke passionately about his desire to make this exhibition of Africa unlike any other. He was committed to probing the dark history of white colonization and putting that history on display. He wanted to radically change negative perceptions of the continent and its people. In the museum world, this felt revolutionary. It felt like a call to symbolic arms that could truly help to decolonize imaginings of Africa. I spent the summer of 2017 working with Bryn Mawr students in the Penn Museum archives, tracking down the history of the African art and artefacts in the Penn Museum collection, and indeed locating the legacy of colonialism and exploitation that haunted much of the collection.

However, the process of radically reinventing the African collection for display tended to offer more challenges than solutions. I've always told my museum studies students that [it] is easier to critique than to construct, and that has certainly [been] born out in this exhibition development process.

How do you distil the continent of Africa into an exhibition, and one legible to sixth graders and college students? There is the sheer complexity of histories and present stories the vast wealth of cultural productions and cultural dynamism, of political and economic and philosophical and gendered histories, not to mention the sheer size and scale of the continent and its people. It is clearly far more than can ever be encompassed in an exhibit on Africa. Representing the continent of Africa in 4,000 square feet is a challenge of epic proportions. And of course, many compromises were made along the way for a variety of practical, political, and ethical reasons ("By putting archival images on display, does it reproduce the original denigration of its subjects?"). However, as we approach the end of exhibition development, I believe we have made substantive interventions in typical representations of Africa. At least we have offered new opportunities for re-thinking [and] re-envisioning as well as critique. Through this, we all collectively, iteratively push for better representations of culture in museums. We'll see (pun intended).

On exhibiting

Monique Scott: You are a distinguished sociologist and scholar of Africa. You also have been quite committed to educating the public through various media including visual media such as television documentaries and museum exhibitions such as the Penn Museum temporary exhibit, *Black Bodies in Propaganda: The Art of the War Poster*[3] to *Tides of Freedom: African presence on the Delaware River* at the Independence Seaport Museum.[4] Why do you think it's important to create these visual stories about African society and politics or culture?

Tukufu Zuberi: I have been studying Africa for 30 years. I have visited a large number of African nations. I have built collaborations with African nations and African universities and large numbers of African scholars around the demography of Africa, around using empirical information to understand and make public policies in Africa, and around how we pursue challenging the dehumanizing narratives which exist about Africa. And I've done that by participating on panels and engaging with groups who are fighting for human rights in various ways but also in making films. I spent 14 years as a "History Detective" on PBS and that gave me a certain perspective about the value and the use of film and video and television as a medium of speaking to a large group of people. When I first found out that I was speaking to millions of people and that I was having these very engaging conversations around race, it really amplified for me the importance of not just writing an article which will be read by maybe a hundred, maybe a thousand scholars but making sure that the information that I was trying to translate could also be received by millions of everyday people, by tens of thousands, hundreds of thousands, millions of families around the country and around the world. I've found that using this audio-visual medium was the most powerful and effective way of doing that and that's why all my exhibitions have a heavy component of audio-visual material, because I try to communicate with people in that way and I try to not hide.

I try not to be a hidden figure behind the exhibit. I want my face there. I want to admit I did this. I had a really big impact on this. You can blame me. You can look at me. You can hear me saying this. To me, that's important because it allows people to both like you and dislike you and that's okay to me.

And as a sociologist it is normal that I would come from censuses – looking at censuses and population information – to going to these other representations of people as they are presented in the media, and in how people look at images that are presented in the media. And [I've been inspired to] not only work in television shows but also make documentaries, and from my documentary film work to the museum. All of these are spaces in which the collective information is the fundamental basis of what you're trying to summarize. With censuses, people can't see the whole census, so you try to give them some summary of it: "this is what it means." And when you're doing a film, you're trying to summarize information so that people can understand what the film is about. And in a museum exhibit, you want to present to people in a voice where they are enriched by the experience of going through the exhibit. And you can do this for various audiences in a very small space and that's what I'm committed to doing.

On *Tides of Freedom*

Monique Scott: Let's talk about one of your exhibitions, the exhibition on slavery at the Seaport Museum. *Tides of Freedom* seemed like somewhat of a radical departure for the institution, one that describes itself as devoted to deepening the understanding, appreciation, and experience of [the] Philadelphia region's waterways. So how did your exhibition find a place there? Was it a comfortable space alongside their more traditional science exhibitions?

Tukufu Zuberi: Several leaders of museums and the curators in Philadelphia had a meeting with me and they wanted to talk to me about putting together a collective exhibit of these various institutions, and to discuss the African presence on the Delaware from the past up until the present. I told them this is a good idea and if you go to the exhibit, you'll see there are key pieces, original pieces, from various museums, from the National Archives, from the African-American Museum, from other museums in Philadelphia. These museums and individuals kind of came together on that. They asked me to tell the narrative and I said let's tell the narrative about something that Philadelphia's high on – first Capitol, the birthplace of freedom, justice, and equality, the Constitution, all of this stuff. And I said let's tell that story in this way because slavery and freedom are key elements of the story.

   If you are going to talk about the waterways, part of the story of the waterways is how profoundly they have been impacted by the presence of these descendants from Africa and these Africans themselves. We went into their archives and we found a ton of stuff. We found a waste book and in the waste book it said something like "five Negro men, so many Negro women so many Negro children and Sally is the owner of these people." Sally gets a name and all of these so-called Negroes get no name. That process of de-naming them, un-naming them is in itself part of the process of dehumanization. And so, this process of dehumanization was a way to talk generally about enslavement and the potential of freedom. So we follow this group all the way up through civil rights as they engage with the waterways, from the woman who ferried people across the Delaware for her master and she lived to be 116; to the arrival of a young lady who was 18 years old in a box where she mailed herself to freedom; to George James Forten and his creating of a shipbuilding company in which he made the sails for boats and fishing nets for boats and had an integrated work force, and was a leading abolitionist in the city; to a steamboat captain today who is shuttling ships back and forth. So the idea was freedom. And we asked each of them [the people in the stories] about how their job, how their task, was related to freedom. Those in the past, we took from what they said. And those in the present, we asked them actually to talk about it. And we used video to express this as well.

Monique: This exhibition probably expanded the minds of many traditional audiences of the Seaport Museum. Tell us a little bit about opening day.

Tukufu Zuberi: Yes. The thing that we did with this exhibit which was very interesting is that we asked this basic question but then we said we're *responsible* for asking this basic question. We can't just ask it as something which is beyond the community. And so, we asked the community to come and allow us to do it . . . We had a Christian leader. We had a Yoruba priest. We had various traditional African religions as practiced in the Americas to celebrate the opening of this journey. I had my name mentioned in the mouth of Sonia Sanchez in her poem to memorialize the event. For us this was a moment in which we were taking and saying *our* voice will be heard. And because our voice will be heard, we will give you [the museum] permission to use it. And so the museum was given that permission. It was supposed to be up for six months and we're now going on six years. I talked to the president of the museum just a couple of days ago and he said [it's] the most popular attraction that they have, that people come to bring classes specifically through this exhibit as a way of talking about African-American history. I just took a class from New York of young African-American girls on a tour through the exhibit. And they can see themselves there because prominently throughout the exhibit are women and what they did on the Delaware and how they contributed to the fight for freedom on the Delaware. So, in some ways the exhibit brought a new audience in the Independence Seaport Museum because it brought that audience who were thirsting for this information to be integrated in the narrative of the museum.

On the African exhibit at the Penn Museum

Monique Scott: You've now moved on to a new curatorial project of tremendous scope renovating the permanent African gallery of the Penn Museum of Archaeology and Anthropology. I can imagine that with that great power comes great responsibility. What made you decide to take on this position? And what do you think you bring uniquely to this traditionally anthropological material as a sociologist? Or with your own background?

Tukufu Zuberi: The Penn Museum is one of the finest examples of the collection of African material culture collected by individuals who sought a certain understanding of Africa. They sought to show that Africans had culture and civilization despite what people were looking at and thinking about Africa – that colonialism was impacting what *was* Africa in a way somehow that what came after colonialism was not so much African. So, to me that's a very interesting conversation because to me that is a colonial mentality. And I am inspired by Chinua Achebe's book, *Things Fall Apart*, because of the conversation he's having about what traditional African culture is and what impact on African traditions is being wrought by the experience of colonialism[5]. How do you talk about African tradition in an era of European colonial domination? That is, in this era of European colonial domination, what is defined as African by those colonizers is part of their hegemonic move to dominate Africa. So, they are thinking of Africa in a way which dehumanizes the African into a colonial subject which somehow being the colonial subject is separate from being the African. And all that is Africa is old and antique and all that is non-African is a result of this kind of impact with the rest of the world.

And to me this myth perpetuates an idea about who Africans are and where Africans can go that renders not only people on the African continent limited in terms of their human scope but all the people who Africa is blamed on producing. And that is black people everywhere in the world whether they're in Brazil, whether you're in the United States, whether in Britain or France or any of these places. Their contribution to the societies that they are in and to human society are always limited by being a lesser contribution to the universal contributions of those who don't need those markers to identify themselves. To be a king is different than being an African king.

Monique Scott: Given all that, weren't you daunted by the task of curating that history into a museum exhibition?

Tukufu Zuberi: Yes, exactly. Because to me it's large, it's major. It's a big thing to do. I've been studying this for a very long time. But I also made sure that I brought in people who could critique me, who could stop me in my tracks and say something. It would provoke me and sometimes I don't go the way they want me to go but I doublethink about how I'm presenting my ideas. And, a lot of times I learn from them about how to move forward. So that's one thing I did to kind of lessen my own fear about moving ahead [with curating] is getting experts who have in their own right a voice to be heard in the direction we take with this project. So, in some ways it is an awesome responsibility because while we are trying to change a narrative and it means that we need to be bold. But we don't need to be hostile on one hand, nor do we need to be foolishly ambitious so that we don't have a basis on which we are speaking. So, I try to start kind of like in my sociological imagination in which every individual biography helps us understand what history is and allows us to translate that history into what we consider to be society. So human society for me stems from the connection of individual experiences into history and how those processes of transition lead to what we call societies of the past. And what we call society of the now.

I've always been very interested in how we construct these notions in the past about Africa, notions from what we think is death, what we think as important, where we think life should be held and who we think is civilized – and civilized enough to consider as part of representing us.

Monique Scott: I've been lucky enough to work with you on the subject and hear your thoughts. Now, I'm wondering what is the motivation you have for this exhibition in terms of breaking stereotypes or challenging pervasive misconceptions about Africa and African people?

Tukufu Zuberi: So here's our kind of underlying theme and kind of the philosophical walking papers, if you will, that we have. We're saying why this object, at this time, in this museum? Because why this object, at this time, in this museum allows us to treat the works that we have as not being static in the sense of [not] only having a significance in terms of the colonial mentality that collecting it, and not only having significance in terms of the craftsman or the artists in Africa who created it, and not only having significance in terms of the cultural traditions which imbue it with meaning, whether those cultural traditions are African-based or in a way European-based – European-based in that they desire these objects to be created for European consumption or in the African base that the object was created for consumption, use, or presentation on the African continent – but to say that that in itself is a human dialogue. It's the human dialogue of the people who do something, the people who create something, and the people who digest/experience that thing which they create. And some of that happens in the African continent, on the African continent within the space of African cultures. But a lot of that happens away from the African continent and that's what all of this is [the museum setting]. We're not in Africa and you do not enter Africa when you enter the Africa galleries. What you enter is a reflection of Africa that represents the choices made by the collectors, by the curators, [and] by the museum administrators. So when we look at the experience of people ruling in Africa and we look at the Benin empire, it is in itself allowing us to peer past the beauty of the material culture of the Benin empire, the royal history, how they governed themselves, their administration, the map of their empire, the fact that they were a military empire.

But we also get to see the crack of that empire by the British Empire which then comes to rule there. So, we get two phases of ruling. While we're having a conversation about statecraft in Africa, the Benin artefacts, plaques, and the altars show the tension in Benin with the Museum experience. Yet the reason we have these pieces of material culture in the Museum is because of the rule of Benin by the British and their "Punitive Expedition." There are layers in the narrative we offer.

Monique Scott: And so, in some ways you're putting colonial practices and collecting and the museum itself on display?

Tukufu Zuberi: It's at the center of what we're displaying. They are there if we allow the objects to represent what they have been. They were something in Africa. But the mental state which allowed someone to write a code on the object – and they all have codes – and to have that code gain meaning and significance and representation in how we identify that object in the museum records, everywhere is on display in our re-design. [This] is part of a process by which we have used this object to represent something. It may have represented Africa in the Bakongo, the Congo village. It may have represented military power, political power, cultural power but the representation of power that it takes in our museum is different. This is because some of that is a reflection of the power of Africa in the first half of the twentieth century. And that powerlessness is represented by taking this object out of context. For a people whose context has been transformed by colonialism. And in not hiding any of that and saying "this is what we're looking at." And all of this mythology that somehow, we have an Egyptian object from 5000 years ago or from 3000 years ago or 1000 years ago that is supposed to reflect Egypt of today is mythology. But that mythology is important only to the extent that we understand that this object was created a thousand years ago and had a journey to end up here. The Great Sphinx ends up here because of both the politics of knowledge and the importance of the object as a historical accomplishment of humanity.

Monique Scott: This is pretty radical for curating an anthropological exhibit.

Tukufu Zuberi: So, I do say this: I take my lead from WEB Du Bois. When he was putting his exhibits together, he was putting his exhibits together not so much to say here is a thing that represents these people, but to say the African was human too and they lived human experiences. They were poor; they were wealthy; they were enslaved and they were un-enslaved; they were free. They were all of these things. And as a consequence, they have a tremendous contribution to make to humanity because of the diversity of what they represent. And so, I take that same approach in looking at these objects from Africa.

Monique Scott: Not just one object, one story object, an object as a complexity, with many stories and histories overlapping?

Tukufu Zuberi: While at the same time engaging the high schoolers, engaging the preschoolers, the young children, as they experience the sheer beauty and tremendous presence of these objects.

Monique Scott: I want to ask this for just your last question: Has this Black Panther moment seemed to help or hurt public education about Africa, and what people might bring to the exhibition?[6]

Tukufu Zuberi: So, the Black Panther moment. I assume you're talking about that moment in the beginning where one of the characters enters the museum and they talk about a Benin object which ends up being a Wakanda object which ends up being part of the metal that is the source of all their power. This is a beautiful moment right now where Benin serves as a central linchpin. There's also this film, *Invasion 1897*, from the Nollywood movement centers around an object in a museum in Europe that belongs to the family of a particular individual and he sues the museum after stealing that object. And so, there are all these different moments in which African objects being in museums are entering the public consciousness. And the space of a museum is becoming a place of performance about the current public consciousness. For example, the music video with Beyoncé and Jay Z where they're going through the Louvre and conclude by being in front of [the] *Mona Lisa*. And what does all of this mean?

The museum's place in public consciousness I think is being explored in unique ways. This is beautiful to me because I see the museum as a very important place for public education, a very important place in which the public receives its facts about what's going on, its representations of the narrative that become the fabric by which the nation state is sustained, and what people think about each other and about others. So, when somebody is not represented or when someone is marginalized in that space or when the only thing we see about African-Americans is enslavement, and the only thing you see about Africans is colonial domination or this precolonial experience which was destroyed by the colonial experience we have a problem. This leads to that [image of] Africa which is torn by wars, torn by poverty, is torn by AIDS, is torn by the lack of industrialization, is torn by high infant mortality. All of these things which seem to represent Africa, in reality the problems that distort Africa's reality. We get an Africa which is disconnected from being Africa, an Africa which is disconnected from the Great Egyptian and Ethiopian empires, for example. It's disconnected from the rise and the splendour of Islam in Morocco or the rise of the empires in Ghana, in the Sudan, in Mali and Songhai. Somehow that gets lost in the story of being part of African history and culture, as being something we could call *pre* Africa or African prehistory or African history before "our Africa." There is no historical continuity like the continuity that you have in other places, in other spaces, which contribute to civilization.

To me we are challenged with the task of addressing these questions for the public because they want to know "Why should I not think of Africa in a bad way?", "Why are these stereotypes wrong?" Well part of what we're doing is we're taking the stereotype and we're flipping it and rather than letting it roast there underneath the savagery which Africa is assumed to have – the mystery which is supposed to be African – we're saying no, Africa is doing the same [as other places across the globe]. Africa is designing like everywhere else, Africa is exchanging goods like everywhere else, Africans are engaged in statecraft like everywhere else, Africa is engaged in spirituality like everywhere else. Africa is part of the human experience. And you can even see this through the blinded eyes of the colonial mentality.

Notes

1 See, for example, Lindfors, Bernth (ed.) (1999). *Africans on Stage: Studies in Ethnographic Show Business*. Bloomington: Indiana University Press or Scott, Monique (2007). *Rethinking Evolution in the Museum: Envisioning African Origins*. London: Routledge.
2 Penn Museum website: African Section Collection Highlights https://www.penn.museum/collections/highlights/african/
3 Penn Museum Website: Black Bodies in Propaganda: the Art of the War Poster https://www.penn.museum/exhibitions/past-exhibitions/262-black-bodies-in-propaganda-the-art-of-the-war-poster
4 Press release from Independence Seaport Museum. Contact Hope Koseff Corse. Independence Seaport Museum Presents Tides of Freedom: African Presence on the Delaware River https://www.phillyseaport.org/images/Tides%20of%20Freedom.pdf
5 Achebe, Chinua (1958). *Things Fall Apart*. London: Heinemann.
6 This is a reference to the 2018 film, *Black Panther*.

3 Science, Islam, and critical thinking

Lessons for Museums

Rana Dajani

Science means using your brain. This is a human trait. It is part of our evolution. It is what makes us human. All humans therefore are scientists. Museums must send this message to people. The scientific method is composed of four simple steps: (1) observation; (2) asking the question, why?; (3) trying to develop an explanation (hypothesis); and (4) testing the explanation to see if it holds true, and repeat if it does not. As humans we apply this way of thinking every day in all walks of life. Not only scientists do this, but artists, designers, writers, even my grandmother. We use this method to improve the fields we work in – to improve life. This is how we connect science with society, and as we have been doing for millennia, this is how we tackle the big challenges facing humans.

Science museums should show how, as humans, we evolved in our thinking. Museums should show how we have used our brains/minds (or not) no matter what our cultural perspective. The thinking mind employing the scientific method is as old as humanity It is part of philosophy and religion. There are many histories of how we used our minds that do not come from a Western narrative. Museums have to show them all. By exhibiting the various ways that we, as humans in different cultures, use our minds, museums unite us, and by doing so, celebrate our diversity.

Traditionally, science museums exhibit the one thread of history that is Western enlightenment. Museums start from the Greeks and Romans and then jump ahead to the Renaissance, leaving the dark ages dark. But in fact, during this time, a whole enlightenment was going on not too far away. Advances in all fields were happening – in physics, medicine, social science, and philosophy.

There have been a couple of travelling exhibitions in European and North American museums that have displayed the history of science as it was developed by Islamic civilization between 700 and about 1400 CE. When the exhibition, Sultans of Science: 1000 Years of Knowledge Rediscovered, toured the Ontario Science Centre in 2009, and then again in 2014, there was a remarkable level of interest from all communities in Toronto. Muslim communities that had not been well-represented among museum visitors came in large numbers and many stayed to visit the rest of the science centre's offerings. A significant number of Muslim families signed on as members and signed their children up for special programmes. The success of this travelling exhibition showed that there is an unmet need for both Muslim and non-Muslim communities to have greater access to exhibits and programs about the influence of Islam on science and technology.

The emphasis on Western civilization creates a bland, inaccurate portrayal of historical facts that is a remnant of colonialism. It implies that anything that does not come from the West or that cannot be measured by the West is not worthy or does not even exist. This is evident in portrayals of Western contributions at the top of museum staircases that then descend to Eastern contributions at the very bottom. Iconic places are given Western names with no effort made to preserve original Indigenous names. For example, who knows what the original name was for Lake Victoria? It was known for millennia by people living on its banks. In the future these travesties in non-representation must be corrected.

The design of museum exhibits shapes the psychology of the museum visitor and imprints upon his/her mind a particular perspective. How can museums help visitors to incorporate their own cultures into their experiences? One suggestion is to allow the museum visitor to engage in a dialogue with the museum exhibits. The exhibits would be displayed in a kind of mosaic with no clear hierarchy. Visitors would be given options to rearrange the exhibits so that they could plan their route during their time in the museum. This could be portrayed digitally on a screen at the entrance for everyone to see, displaying the multiple awarenesses of the world and other ways of thinking to the public in a continuous dialogue. These data could be used later by the curators as they develop new exhibits or update old ones. In the future, how exhibits are arranged spatially within the museum would be transformed by drawing on new sets of rules that consider the visitors' feedback. Museums should create exhibits that trace the multiple threads of history, taking the visitors on a journey of discovery. That will allow us to learn lessons from life to forge a better future.

There are 1.6 billion Muslims in the world today, and they hail from different cultures and countries. Islam is not bound by ethnicity, kinship, geography, history, or language. Muslims and Islam have been a major contributor to science around the world throughout history. Science museums should exhibit the philosophy and history of how the thinking mind developed in such diversity. This will pay tribute to all histories, will be fair in representation, and will allow the world to learn lessons from the past that will benefit it going forward into the future.

Consider the subject of biological evolution. Through its texts, Islam preaches and mandates that we use our minds to observe, contemplate, question, and test our ideas. There are multiple verses in the Quran that call for contemplation, such as, "Indeed, in the creation of the heavens and the earth and the alternation of the night and the day are signs for those of understanding" (Quran 3:190). This verse shows that we are strongly encouraged, even required, to *think* about the creations of Allah. This would, by natural course of human action, support the practice of scientific thought, where we inquire further into the workings of nature in order to understand it. Therefore, there is no conflict between Islam and science.

Throughout history Muslims and non-Muslims living within Islamic civilizations adopted this approach. Caliph al Mamun said that reason and faith can coexist and that by fully opening the mind and unleashing human creativity many wonders, including peace, are possible. It did not matter whether you were female or male or from which religion you came. The first university in the world was built by a Muslim woman, Fatima Alfiri, but we don't hear about that in today's museums. Al-Qarawiyyin University was founded in 859 AD in Morocco, more than 200 years before the first European university was established in Bologna, Italy. Al-Qarawiyyin University is still operating today: Pope Sylvester II studied there (ca 1000) and was considered to be one of the most enlightened popes. The Jewish scholar Maimonides (twelfth century) also studied at this university.

Islamic civilization has achieved many advances in science, humanities, and the arts. For example, the first person to realize that light enters the eye, rather than being emitting from it, was the tenth century Muslim mathematician, astronomer, and physicist, Ibn al-Haitham. He invented the first pin-hole camera after noticing how light came through a hole in window shutters. The crank-shaft, a device that translates rotary into linear motion and is central to much of the machinery in the modern world, was created by an ingenious Muslim engineer, al-Jazari, for raising water for irrigation. Many modern surgical instruments have the same design as those developed in the tenth century by the Muslim surgeon, al-Zahrawi. His scalpels, bone saws, forceps, fine scissors for eye surgery, and many of the other 200 instruments he devised are recognizable to a modern surgeon.

The development of mathematics by Arab scholars, and by other peoples, many of them Islamic and living in the Mediterranean area and in Mid- and South Asia, is a singular contribution to the progress of abstract thought – a history not known by most people living in Western societies. The origin of the concept of zero is an interesting example. Mathematics in general, and all the practical uses of mathematics, require the concept of zero. The earliest mention seems to come from Sumeria via Babylonia (arguably in the lineage of Arab scholars who continued to work on mathematics throughout history until today). Further development came from India starting four to five thousand years ago and lasting over a period of about 3,000 years. And the concept was developed independently by Mayan civilizations about 1700 years ago. It was also an Arab scholar, Alkhowarizimi, who used the concept of zero in algorithms that today are the basis of the internet.

Open-minded and critical thinking, applicable in every time and space, were hallmarks of Islam. If you are not critical, you are not a good Muslim. On the contrary, you are rewarded for being critical even if wrong: "When a judge gives judgment and strives to know a ruling (*ijtahada*) and is correct, he has two rewards. If he gives judgment and strives to know a ruling, but is wrong, he has one reward."[1] Other verses in the Quran encourage critical thinking and question even God: "And [mention] when Abraham said, 'My Lord, show me how You give life to the dead.' [Allah] said, 'Have you not believed?' He said, 'Yes, but [I ask] only that my heart may be satisfied.' [Allah] said, 'Take four birds and commit them to yourself. Then [after slaughtering them] put on each hill a portion of them; then call them – they will come [flying] to you in haste. And know that Allah is Exalted in Might and Wise.'" (Quran: 2:260).

This verse invokes two points: 1) asking questions, even of Allah, is allowed, and 2) the verse invokes the scientific method, i.e. God shows Abraham through evidence how He resurrected the dead birds. meaning that to answer any question you need to show evidence and not rely on blind belief. Ie God did not tell Abraham to believe blindly, but on the contrary He showed him concrete evidence. The answer by God indicates a broad understanding of the intertwined quality of life and of our total ecology. The dead bird parts left on the hills give rise to living birds – a joyful portrait of the cycle of life that has divine roots, but that can be questioned and examined for further details. This is a demand to find answers in the world around us. The answers are in the world itself for the scientist to explore. The verses spell out the scientific method that starts by observing nature, then poses questions, and lastly attempts to find answers. And posing questions also requires the consideration of different possible solutions. The scientist and polymath Alrazi, in his preface to his Book of Secrets (ca 900), gives a model for scientific critique that says one must take seriously the opponent's proposition before engaging in a debate.

There were Islamic scholars who, through their observations, attempted to explain the diversity of living organisms around them. Al Jahez (ca eighth and nineth centuries) discusses rudimentary theories of evolution in his book, *The Animal*. Later, Ikhwan Alsafa and Alrazi went further by suggesting that humans changed over time. Al Rumi described his theory of evolution (although not precisely Darwinian) in a poem titled "The Progress of Man".[2] The poem describes the emergence of humans from inanimate matter through a series of stages from inanimate to plants to animals. He also proposes that humans will change as well in the future. As can be seen in his poem, Al Rumi describes evolution beautifully. He indicates that different creatures came from other creatures and not all were created spontaneously, ideas that are very close to Darwinian thinking but in the context of the scientific knowledge at that time. Islam encourages and even requires that humans embark on a journey of discovery with total freedom to interpret the world. These scholars were encouraged to question and research, as part of what Islam is.

This seems contradictory to the situation in the twentieth century where research points to Muslims rejecting evolution. So, what happened? When Darwin published his book *On the Origin of Species* in 1859, it was read by Muslim scholars, such as Aljisr in the Levant area, and it was accepted and discussed. In contrast, Christian scholars in the Levant area, such as those studying at the American University of Beirut, followed in the footsteps of their predecessors in Europe and opposed Darwin's ideas. But at the same time, in that period, starting in the late nineteenth century and into the twentieth century, the Ottoman Empire was waning, there was less education, and there were few scientists in the Islamic world who had the expertise to understand Darwin. Muslim religious scholars began to adopt the Levant Christian stand that rejected Darwin. And colonialism was flourishing; England and France had appropriated the Levant and other parts of the Muslim world. The colonizers used *social* Darwinism as an excuse to extend their influence over other people. Darwinism became even more reviled through its association in the minds of the colonized with oppression and injustice and was in turn demonized. Therefore, a mixture of ignorance, colonization, and Christian religion resulted in the rejection of the theory of evolution by Muslims in the twentieth century.[3]

As Muslim countries gained their independence, and education became available to all, Muslim scientists realized that they had no conflict with evolution. Later, when researchers studied how Muslims perceive evolution around the world, they used terminology adopted by the West. When translated within an Islamic context, the word "creationism" has different meanings: creationism means that a Muslim believes that there is a god, but not necessarily that He created things spontaneously. In the Christian Western context, creationism means that God created everything spontaneously. Terminology has led researchers astray and has resulted in erroneous conclusions about Muslim beliefs about evolution.

We must be aware that others have different perspectives, and that there are differences in conceptual issues as basic as the relationship between religion and science or even the role of humans in the universe. Different cultures perceive the position of *Homo sapiens* in different ways: Western cultures believe that the world exists to serve humans. Other cultures, such as Islam, believe that humans exist to serve the world, to take care of it, and to live in harmony. Both approaches should be presented in museum exhibits, so that visitors can begin to understand the various mind-sets of different communities and how they have come to understand the world around them. The importance of critical thinking is to realize that there is no one true mind-set or belief – one has to be aware and critical of the various mind-sets – of how our roles in the universe are perceived, and how we interpret history. One cannot use only one lens to look at the past for interpretation.

It is the duty of science museums to encourage curiosity, critical thinking, and doubt for the sake of exploration. We must not convey the tone that all the science we know today gives us facts that are unchangeable forever. We must, as all good scientists, be humble and indicate that for today, within the context of the present, these are the best explanations of the phenomena around us. And our ideas may change if circumstances change or new advanced tools are developed. Museums, too, must keep an open mind, or they will oppress minds by insinuating that all questions have already been answered.

Museums need to send the message that we encourage humans to use their minds to question, and not accept anything less. This is a call to action to keep searching for the truth, so our young emerge from museum experiences inspired to be the bold explorers of the frontiers of science, free of inhibitions and with open minds. By sharing the lessons of other cultures, we free our visitors from the hegemony of any culture, giving them the opportunity to trust themselves and to go forward with pride, tempered with humility and respect for others.

These are great examples that teach lessons to visitors both young and old on how to resolve miscommunication and misunderstanding and create healthy discussions and debates that will advance science rather than stifle it. An example is the story of climate change. It took a long time for investigative journalists to bring this issue forward and dispel the myths. Some scientists were coerced by companies who would use their resources to downplay the evidence that climate change is real. Science museums should tell the stories in their exhibits of the brave scientists who were ethical and stood up for justice whatever the cost. Children are always looking for heroes to emulate. Showing these scientists as heroes in these stories will enable them to be role models that will inspire young people.

Museums should showcase the roles of scientists by presenting examples of what was done wrong and how it should be done in the future. Visitors need to be given the chance to think of what they could do to boost social responsibility in science. Museums should indicate that this is an ongoing process that never ends. Decisions we make today that appear to be ethical may change and will be challenged with every new discovery in science. With each new discovery, the consensus will be challenged, and that is the journey of life that never ends and what makes it wonderful. Museums in this context should point to the role of science in upholding justice – how to do what is right for humans.

Social responsibility in science can have at times important political dimensions. In the past it was often scientists who championed the boycott of institutions in South Africa in the fight to stop the practice of Apartheid. The conjunction of science, social responsibility, and politics can be seen today in the case of Palestine and Israel. Stephen Hawking, the late famous physicist from Cambridge University, supported the boycott of Israel by Palestinian academics by refusing to attend a 2013 conference on the future of Israel. Museums should tell the stories of how scientists have played a role in politics and the important debates in today's complex world. This is important for another reason. When scientists are involved in politics, they make sure that science is not taken out of context and used by politicians to advance their own agendas. In the past, as can be seen when studying the history of anthropology and human biology, science was used inaccurately to push forward racist agendas. It is the responsibility of scientists to be vocal about this and to lead initiatives to hold politicians accountable: Every human being is responsible to use the knowledge they have gained to make the world a better place.

Museums can learn from education pedagogies in this context. There are many pedagogical approaches that encourage students to learn better and gain a deeper understanding of the subject matter, and that also encourage them to employ what they have learned in their communities. The objective of academic study should be to create citizens who can identify problems, develop solutions, and become changemakers in their communities. Museums should have the same goals. For example, museums can employ service learning that links personal and social development with academic and cognitive development. Eyler and Giles[4] summarize their observations by saying that in the service-learning model, "experience enhances understanding; understanding leads to more effective action." Science museums can educate their visitors by providing service-learning opportunities that bridge experiences gained from the museum with local communities that would benefit. By sharing their experiences, visitors learn how to employ the knowledge/skills they have gained to help their community. Research has shown that people get a good feeling from helping others because it is part of who we are as humans.[5]

Media can also be harnessed by museums to get the attention of visitors and to spread the museums message. Museums can invite their visitors to think about how to explain scientific concepts to the public through various media outlets. Through service-learning, museums can collaborate with schools and other educational institutions in new ways. Collectively this creates a sense of responsibility towards the community and encourages civic engagement. The added advantage is that through science museums, the concept of civic engagement can be extended to the entire globe because of the universality of science. This is similar to the concept put forward by Hakan Altinay on global civics: "Our world is becoming more interdependent. What happens in one part of the world affects lives in other parts. CO_2 emissions, infections, financial integration, nuclear leaks and several similar dynamics have significant global consequences. These dynamics are pushing us together and intermixing our destinies. Our increasing global interdependence means, among other things, that our lives are no longer solely authored by us, but are being co-authored with others. In order to navigate our increasing interdependence, we need global civics. We need to decide what sort of rapport we wish to have with the billions of others with whom we share our planet and destinies, but not our citizenship."[6]

Citizens of all nations are residents of an ever-shrinking village. With the acceleration of globalization and its effects, cooperation at all levels of engagement is more important than ever. Kofi Annan, former Secretary General of the United Nations said, in reference to global civics, the global community, and the interconnectivity of humanity: "I hope that universities and think tanks around the world will deploy their significant reservoirs of knowledge and creativity to develop platforms to enable students to study and debate these issues."[7] Global civics is understood as a social contract between the world's citizens in an age of interdependence and interaction. The term was first developed by Dr. Hakan Altinay, a World Fellow at Yale, a non-resident senior fellow at the Brookings Institution, and a global ethics fellow at the Carnegie Council for Ethics in International Affairs. Global civics is a response to our increasing global interdependence as well as a search for how to manage that interdependence. It highlights the importance of uplifting our perspectives towards global interdependence and its attendant responsibilities, including environmental issues. We have developed a course on global civics from the Science/Muslim perspective for undergraduate science students in the Arab world. According to the Prophet Mohammad, "Everyone is a guardian."

The task is enormous: science will play a pivotal role, bridging human and environmental concerns and providing key clues on how to overcome the challenges. Museums should create paths for visitors to understand history and give freedom for each person to create their own journey. As Pulitzer prize winner Marilynne Robinson, whom I met while at the Radcliffe institute, said during a talk she gave, to feel comfortable in doubt. To do this practically, one must design museums where the audience actually participates in the creation of the experience, so that the community and the individual are engaged and start creating the museum themselves. In this way, visitors will feel that they belong and are welcome in the museum because their voices are part of the experience and they see how the museum can serve them in their everyday life. Museums in this way can serve to create communities of understanding and communication when everyone sees himself inside the museum and the relationship is ever changing as we ourselves change.

Notes

1 Hadith of *Bukhari*, 9.133.
2 Guessoum, N. (2011). *Islam's Quantum Question: Reconciling Muslim Tradition and Modern Science*. London: I.B.Tauris.
3 Stott, R. (2013). *Darwin's Ghosts: The Secret History of Evolution*. New York: Spiegel & Grau. Guessoum, N. (2011). *Islam's Quantum Question: Reconciling Muslim Tradition and Modern Science*. London: I.B. Tauris.
4 Eyler, J., and Giles, D. (1999). *Where's the Learning in Service-Learning?* San Francisco, NJ: Wiley.
5 Buchanan, K., and Bardi, A. (2010). Acts of kindness and acts of novelty affect life satisfaction. *The Journal of Social Psychology*, 150(3): 235–237. Aknin, L. B., Dunn, E. W., and Norton, M. I. (2012). Happiness runs in a circular motion: Evidence for a positive feedback loop between prosocial spending and happiness. *Journal of Happiness Studies*, 13(2): 347–355.
6 Altinay, H. (2011). *Global Civics: Responsibilities and Rights in an Interdependent World*. Washington, DC: Brookings Institution Press.
7 As cited in Altinay, H. (2011). *Global Civics: Responsibilities and Rights in an Interdependent World*. Washington, DC: Brookings Institution Press.

4 Museum mantras, teachings from Indian country

Posterity is *now*; failure *is* an option, and repatriation is a *foundation* for research

Jen Shannon

The American Alliance of Museums has concluded that museums are the most trusted source of information in America.[1] This is a big responsibility; how do we manage that, and what do we do with it? I have asked that question from the point of view of someone who has studied, worked at, and taught about the intersection of Indigenous peoples and museums since 1999. Over the years, I have gathered up mantras that help guide my practice.[2] They are all informed by a fundamental truth that W. Richard West, Jr. (Cheyenne), former director of the Smithsonian's National Museum of the American Indian, once said while I was working there: "We love museums because they have our stuff; we hate museums because they have our stuff."[3] Therein lies both acknowledgment of a traumatic and troubled past and hope for working together in a place that is valued, even if in different ways, by both museum professionals and Indigenous community members.

That hope comes in the face of a difficult history between science museums and Indigenous peoples, which I will not go into detail about here. But it is something that each of us should educate ourselves about, especially if we work in institutions with collections that originate from Indigenous peoples, because science museums played a part in the oppression, perceived vanishing, and objectification of them. Scientific racism, colonial and extractive collecting practices, the disinterment of Indigenous people's ancestors for scientific study, the treatment of Indigenous individuals as specimens, "frozen in time" representations, and the mere fact that Indigenous peoples' material culture and bones are housed in *natural* history museums are the context in which we work and attempt to build relations of trust. As museums embrace the turn to social justice and community engagement, I believe that they should be in part guided by peoples whose items they house in their collections.

At the heart of these mantras, these teachings from Indian country, is a call to reframe how museums think about Native peoples, collaboration, and repatriation – a call to reimagine the museum and its purpose. We want to embrace a different value system, one that welcomes other ways of knowing, reconsiders what counts as sources of expert knowledge, and engages with Indigenous individuals as partners, interlocutors who help us to reorient our thinking and our understanding of the purpose of museums.

This reorientation has been going on in scholarship and in the everyday practice of collections, visits and consultations with Native peoples in the United States for some time. Native peoples, anthropologists, and museum professionals have all influenced a transformation in museology in the United States to embrace this perspective. Change has been driven by Native peoples' activism, critical scholarship, and interactions with us during our fieldwork; anthropologists' critiques of colonialism and representation; the Native American Graves Protection and Repatriation Act (1990) that mandates consultation between museums and US Tribes; and our embodied practice working with Native American items in our care, shaped through consultations and instructions in proper care from an Indigenous perspective.[4] Accordingly, our ways of seeing and relating to the items in our care has changed and consequently so has our understanding about the potential, purpose, and practice of museums with anthropology collections.

Posterity is *now*

This is the mantra I repeat to myself and others, including my students, most frequently these days. It was inspired by a quote I read by Robert Janes, former director of the Glenbow Museum that had produced a groundbreaking collaborative exhibition with the Blackfoot Nation. As Janes explains in his reflection on the repatriation of medicine bundles to Blackfoot community members: "The museum profession is fond of saying that 'museums keep things for posterity.' By 1998, we had concluded that *posterity had arrived* – both for the Blackfoot and for the Glenbow."[5] This is a radical statement. It fundamentally calls into question assumptions about who and what museums are for. Posterity is not (or not only) the general public in an undefined future: it is Indigenous peoples, today. This forces us to address and redress the history of relations between museums and Indigenous peoples, and it implores us to redefine the museum's purpose.

Taking Indigenous ways of knowing seriously, we come to understand that heritage work is not just about connecting to the past; it is also about maintaining the health and well-being of communities in the present and the future.[6] Exemplary projects include collections visits focused on language recovery; digital web portals that provide access to museum collections and photographic archives; collaborations between museums, Indigenous communities, and medical professionals focused on the role of heritage in well-being; interaction with collection items to foster relationships between isolated elders in urban areas; exhibitions that highlight projects promoting language and sovereignty; and long-term loan programs that bring items from museum collections into communities. According to Karl Duncan, director of the Poeh Cultural Center that has a long-term loan of one hundred pots from the National Museum of the American Indian, bringing these items into the community helps to "bring back these traditions, bring back these reminders, they'll enable Native people, enable Native communities, to be healthier, to be happier. To solve a lot of the problems that we have going on today."[7]

In my work with members of the Mandan Hidatsa Arikara (MHA) Nation, which began with consultation about a collection in our museum in 2011, their guidance of our research has led to a series of community-directed projects aimed towards creating the conditions for community well-being. These include tribal members reclaiming sacred items from the museum to enable contemporary spiritual practices, elders using video to recount the history of relocation in their community and to explain to young people how that terrible time feels similar to the current oil boom, missionary family members sharing previously inaccessible photographs of tribal ancestors and relatives, tribal filmmakers co-facilitating video workshops on the reservation so that community members can share their own stories in the midst of an oversaturated and negative media environment, and, through the recovery of a previously unknown archival video, tribal members hearing commemorated leaders speak across the generations in their native language.

Posterity is Now – it is Indigenous peoples, today. They are demanding that we share authority in the care and interpretation of their material culture, and they are inviting those of us who work in museums to participate in their efforts to increase the health and well-being in their communities. But these lessons go beyond their communities alone. We can surely imagine this focus on collaboration, connection to heritage, and community well-being in relation to other communities that museums serve. So, can we reimagine the museum, alongside Native community members, as a place that enables relations and practices that support community health and well-being? And museum professionals as allies in doing so? What would it mean for your museum to reorient your notion of posterity – to now?

Failure *is* an option

I am a big fan of space science history, so it's not surprising that this mantra arose from the phrase, "Failure is not an option," which was popularized in the film *Apollo 13*. It underlined the lethal consequences should the engineers' calculations about the endangered mission fail. It is a rallying cry for high-stakes, goal-directed work. But in collaborative work with communities, where the goal is the maintenance of an appropriate process or relationship, the opposite is true: failure *is* an option – it has to be. In fact, it is a crucial component to the success of the mission.

I am not referring here to the sense of "failing forward," an idea highlighted in the American Alliance of Museums (AAM) Trendswatch 2017 section titled, "Failure, Falling Toward Success." That is more about risk-taking being essential to innovation and creativity. Collaboration is about shared authority, trust, and an openness to other ways of knowing: essential ingredients to working with communities that have been historically underserved by museums, or in the case of Indigenous peoples, undermined by them.

In my role as curator I practice collaborative anthropology, in which Indigenous peoples participate in setting the goals, methods, and outcomes of research. This helps to ensure proper care of collections, representations that are not stereotypical, and it engenders mutual respect. I have labelled my work that embraces this mantra "tentative anthropology."[8] Tentative is defined as experimental, unsure, uncertain – not definite, but hesitant: What shall we do together? We don't have to do anything. We can stop or change direction at any time. What would be meaningful to you? How should we proceed? Therefore, tentative anthropology is predicated on the risk of possible failure; the fact that the choice is out of our hands is what matters. It also means, in complying with community wishes, not including that pivotal story in my journal article, or not displaying that item my student selected to feature in a museum exhibit.[9]

Collaborative work often moves in unanticipated directions, and Failure is an Option suggests we be prepared to do so. I was greatly surprised where our research led with the MHA Nation over the last seven years: from a repatriation consultation regarding MHA items in the museum to an oral history documentary about the donor of the collection. And then, we were invited to produce a sequel documentary about the oil boom in North Dakota. We arrived a couple of years later to do so, with funding in place, but by then the community had been inundated with documentary teams telling negative stories about their community and the boom. Although they did not say it outright, the lack of interest and participation suggested the second documentary was a failure. We accepted that and did not insist that the project move forward. Instead, we changed direction and co-facilitated, with local MHA filmmakers, a series of video workshops for community members to create and share their own stories. We were able to amend the grant to support this new initiative. We held community-wide film screenings and hosted the short films on a project website and, at the request of workshop participants, on Facebook. One video, about speaking the Hidatsa language, was viewed over 3000 times.

Failure is an Option insists that we share authority with the communities we serve, especially those whose items are in our care. So, can we reimagine the museum, alongside Native community members, as a place that values their authority, concerns, and ways of knowing? Can we see risking failure as fundamental to productive relationship building? What would it take, what resources and philosophies must be in place, to enable your museum and its staff to be flexible and adaptive to the emergent process of working collaboratively with communities?

Repatriation is a *foundation* for research

If we take seriously the idea that museums must reorient their purpose – or at least embrace as one of their purposes – to supporting the present and futures of Indigenous peoples, then an excellent way to demonstrate that is through consultation and repatriation. When the Native American Graves Protection and Repatriation Act (NAGPRA) passed in 1990, many people imagined that collections would be emptied and research would cease. My experience has been the opposite: repatriation has been an excellent foundation for research, led to an increase in knowledge about our collections, and created a more positive and meaningful experience in the museum for both Indigenous peoples and museum staff.

The series of MHA Nation projects mentioned previously began with a NAGPRA consultation in 2011 between our museum and MHA Nation tribal liaisons. Our proactive reaching out to consult, and the good experience of our first meeting, led to a sense of building trust and a desire to keep working together. The various research projects we did together began before the sacred items were returned in 2014 and continue today in new and unanticipated directions: we are now brainstorming about training programs for interns in their new interpretive centre and a comic book about the history of repatriation in their community.

Another example is our work with the Navajo Nation. We contacted them in 2009 to inform them of medicine bundles in our collection and invited them to consult. In January 2010, with their support, we applied for a National NAGPRA grant to facilitate their visit to our museum. By March of the same year, we also submitted an American Alliance of Museums (AAM) Museums Connect grant to work with Navajo Nation members on a research project called iShare. In 2011, the NAGPRA liaisons came for the consultation and insisted that we photograph all of the items and enter them into our collections management system before they were repatriated.[10] They provided details on the names and meanings of each item for our records and recorded on video answers to students' questions about repatriation for teaching purposes. At the same time, our collaborative AAM *iShare* project with the Navajo Nation Museum was underway, in which Navajo and Paiwan, Indigenous peoples from Taiwan, travelled to each other's homelands, co-produced a collaborative website, developed culturally informative teaching kits about themselves to share in each other's schools, and gave public talks at our museum in Colorado.[11]

Repatriation is a Foundation for Research reminds us that consultation and repatriation are a form of restorative justice that, although it cannot erase the past, signals to Indigenous communities that we are willing to do something about it – and include them in the process. It often creates the conditions for building trust and mutually beneficial relationships and therefore reorients, in community members' minds, what might be the purpose of museums. While NAGPRA only applies to federally funded institutions and federally recognized tribes in the US, the spirit of the law can be adopted and applied anywhere. There are plenty of examples of international consultations and repatriations.

So, can we reimagine the museum, alongside Native community members, as a place that addresses past injustices, and endeavours to build new and more equal relations with Indigenous peoples? Can we see returning Native ancestors/human remains and sacred items as, rather than "emptying museums," actually bringing more knowledge, and sometimes different kinds of items, into the museum? Can we reorient our understanding of repatriation so that it is defined as a way to build relationships that endure and enhance the research we do, as a means to create the conditions for more meaningful experiences for museum staff, originating communities, and the public? How might you explain repatriation to the staff of your museum so that it can be seen as a way to forge new relations, engage in collaborative research, and move in unanticipated directions?

Anthropology and the science museum

I am not advocating for a way for museums to work with Indigenous peoples – well, not only that; I am suggesting that we learn from what they are teaching us and incorporate that more broadly into how we work with everyone. I am suggesting that anthropology be not just a collection in science museums, but also an approach to the world, people, and collections that informs our philosophy and our practice as museum professionals.

The anthropological approach has more often than not brought us out of the museum and into communities, and through their demands for reciprocity has pushed us to produce materials that are more accessible to them. Here I do not mean finding new ways to bring underserved communities into the museum or make label text more approachable, both of which are important. I am referring to finding ways distinct from exhibits or programs to send information and the results of our research into communities through media that meets them where they are at. For example, our collaborative research has been featured in tribal newspapers and radio, DVDs that were distributed around the community, websites, and even a comic book.[12] For the oral history documentary we made with members of the MHA Nation, there were two versions: one for the general public, and one for the community that included fifteen extra minutes focused on our museum, its collection, and the 2014 repatriation of sacred items to the tribe. The idea was to let community members know, in a medium they requested and are familiar with, what our museum holds and that we welcome them to visit. These projects that arose from a consultation in our museum engaged our museum and its staff in supporting Indigenous communities' well-being.

On a broader scale, we are increasingly being urged to see the museum as having a role to play in our wider society's well-being, too. At this time, there is concern about a growing lack of empathy in American society – and many professionals in museums see our institutions as a way to provide opportunities to increase empathy, with originating communities and with the public.[13] The AAM highlights the power of museums to close "the empathy deficit."[14] Anthropology is a discipline whose methods and approach are all about creating empathy: we seek, through participant observation and interviews and collaborative research methods, the local or native point of view. Increasing tolerance and empathy is fundamental to our disciplinary contribution. So, this essay is also a call for more integration of anthropology into science museums.[15] In addition, if we take the notion of the Anthropocene seriously, we can no longer separate natural from human history, or Indigenous peoples from non-Indigenous peoples, as we are all part of coevolving sociopolitical and ecological systems.

As I came to the end of writing this essay and thought about the purpose of this book, it occurred to me that including more voices, even unheard voices, is not enough. Perhaps the mantra of multivocality has done its work in museums, and a new way of thinking about inclusion needs to be embraced. Museum anthropology with Indigenous peoples has much to share from decades of experience learning to focus more on collaboration and trust, less so on inclusion of "voices." Inviting people into dialogue is not enough. Invite people to help shape the process rather than only representing their voices or point of view in the product. Being supportive of their goals, sharing authority, and establishing a foundation of trust and reciprocity is key. Being open to other ways of seeing and knowing the world, including other ways of seeing and knowing the museum and its purpose, would be ideal.

Notes

1 American Alliance of Museums. 2019. "Museum Facts & Data." https://www.aam-us.org/programs/about-museums/museum-facts-data/ Accessed 7/16/2019.

2 The mantras in this essay are drawn from my own experiences and previous publications. For a deeper dive into these ideas beyond the citations in the notes below, see *Our Lives: Collaboration, Native Voice, and the Making of the National Museum of the American Indian.* Santa Fe, NM: SAR Press, 2014. For more detail about the projects I mention here see http://spot.colorado.edu/~jshannon/. I have highlighted repatriation as a foundation for research in a number of talks since 2010, the most recent being "Allies in Unexpected Places: Transforming museology through partnerships between Indigenous peoples, anthropologists, and museums," Oxford University Pitt Rivers Museum, April 2017.

3 This is my recollection of hearing him say this phrase. See also Cobb, Amanda (2005). Interview with W. Richard West, director, national museum of the American Indian. *American Indian Quarterly*, 29(3–4): 519 in which he states something similar.

4 For more on how everyday practice informs these perspectives, see Krmpotich, Cara (2015). Teaching collections management anthropologically. *Museum Anthropology*, 38(2): 112–122 and, Shannon, Jennifer (2017). Collections care informed by native American perspectives: Teaching the next generation. *Collections: A Journal for Museum and Archives Professionals*, 13(1): 205–224, 2018.

5 Janes, Robert (2015). The blackfoot repatriation: A personal epilogue. In *We Are Coming Home: Repatriation and the Restoration of Blackfoot Cultural Confidence*, edited by Gerald T. Conaty, 241–262, 255. Edmonton: Athabasca University Press., emphasis mine.

6 The World Health Organization's definition of well-being is the condition "in which every individual realizes his or her own potential, can cope with the normal stresses of life, can work productively and fruitfully, and is able to make a contribution to her or his community;" see www.who.int/features/factfiles/mental_health/en/, accessed 2/22/2018.

7 From "Coming Home Project," at www.youtube.com/watch?v=4ASFgGofBeU, accessed 3/9/2018.

8 Shannon, Jennifer (2017). On being a tentative anthropologist: Collaborative anthropological research with indigenous peoples in North America. In *Practicing Ethnography: A Student Guide to Method and Methodology*, edited by Lynda Mannik and Karen McGarry, 58–65. Toronto: University of Toronto Press.

9 In "Posterity is Now," a manuscript under review, I explain why the central motivating story of the essay was removed at the request of a community member and instead focus on the importance of abiding by that request as a fundamental part of collaborative research and tentative anthropology. See also "Questions in Culture: Research Revealed," a 2018 exhibit at the University of Colorado Museum of Natural History in which, after a student consulted with a Hopi cultural specialist, she removed an item from display and replaced it

with a sign: "Culturally Sensitive Item – Not Appropriate for Public Viewing" (photographs available at https://jenshannonanthro.weebly.com/publications-presentations--exhibits.html).

10 The images are restricted to those who obtain permission from the Navajo Tribal Historical Preservation Office. For more information, see Shannon, Jennifer (2017). Collections Care Informed by Native American Perspectives: Teaching the Next Generation. *Collections: A Journal for Museum and Archives Professionals*, 13(1): 205–224, 2018.

11 See Shannon, Jennifer (2014). Projectishare.com: Sharing our past, collecting for the future. In *Museum as Process: Translating Local and Global Knowledges*, edited by Ray Silverman. New York: Routledge. The iShare website is available at http://en.projectishare.com/, accessed 3/28/2018.

12 See Shannon, Jennifer (2014). *"My Cry Gets Up to My Throat": Reflections on Reverend Case, the Garrison Dam, and the North Dakota Oil Boom*, documentary film 60min tribal and 47min public cut, available at http://vimeo.com/118052260. Atalay, Sonya, Jennifer Shannon, and Swogger, John (2017). *NAGPRA Comics Issue 1: Journeys to Complete the Work: Stories About Repatriations and Changing the Way We Bring Native American Ancestors Home*, available at https://drive.google.com/file/d/1nGY5U0P4P0XKXlJa9J5H63bJwb7P9eMS/view?usp=sharing.

13 In discussing the decline in American empathy, AAM, Scientific American blog, and others often refer to the 2011 study by Konrath, O'Brien and Hsing, C. Changes in dispositional empathy in American college students over time: A meta-analysis. *Personality and Social Psychology Review*, 15(2): 180–198.

14 See AAM Trendswatch 2017, pages 8–14, available at https://aam-us.org/docs/default-source/center-for-the-future-of-museums/trendswatch-2017.pdf, accessed 3/28/2018.

15 Anthropologists have also studied science museums as institutions of power and knowledge production; see for example MacDonald, Sharon (2002). *Behind the Scenes at the Science Museum*. Oxford: Berg Press about the Science Museum in London; and Diana Marsh's more recent book currently in press with Routledge about the production of paleontology exhibitions at the Smithsonian Institution National Museum of Natural History.

5 Ensuring our people have a place that feels like home

Elijah Benson, Royce Young Wolf, and Mary Baker Price, with Jen Shannon

We are members of the Mandan Hidatsa Arikara (MHA) Nation, which is located in North Dakota. We gathered at the 2018 International Conference of Indigenous Archives, Libraries, and Museums (ATALM) to have a conversation about the future of museums and the future of our own MHA Interpretive Center, which is currently under construction. What follows is our thoughts about our experiences in museums, and what we want to create in the MHA Nation's new Interpretive Center: a place that feels like home, welcomes community members from on and off the reservation, and serves as a place to rebuild our cultural knowledge and language to ensure our vitality into the future.[1]

Royce Young Wolf:	So, Jen's writing a book chapter, and the book's editors want to do a complimentary chapter with the up-and-coming generation of people being involved in interpretive centres, museums, and what motivates them, and what they would like to see, what they want to do different –
Elijah Benson:	– and want to make it better. Well, for me, I found it very hard to access museums, it was very foreign for me. I wanted to learn about specifically where I'm from. Individuals in the museum had this type of "This is mine" attitude, without ever thinking that this is where I come from, these materials are actually *of me*. And so that was the thing that had the biggest effect on me emotionally. Of course, reconnecting to those materials was very empowering for me, for sure. But it did strike up many emotions. The biggest emotion for me has been anger, out of that experience, out of that space. It's not right. And having been disconnected – going to the city and being totally disconnected from my cultural heritage and the writings, photographs, artefacts, and so forth – having

been disconnected from that, and going back into these spaces, like the archives and the history centres, and falling victim to that disconnection again – that was the most clashes I had, for sure. About this disconnect – for me, what it comes down to is this form of psychology of my dad's generation. How I've come to understand it, there's been a lot of what to think: You've got to go to school, or you've got to go to work 9-to-5, you've got to work hard, you've got to bust ass. You know, cowboy life, like that. But now, what I've seen in my generation is that there's more of a question of what you want to do. What do you want, what makes you *you*, what makes you happy? That's what I've gathered in the conversations with my dad and my uncles and my aunts. With my generation coming up, we're hungry for these stories, language, ceremonies, and so forth.

I wasn't on the rez other than going for a weekend, for that devastation of losing a family member, or just a week for whatever kind of community event was happening. I didn't know the language and other cultural aspects that my grandma had, and her mom and grandparents, and so forth. This knowledge is something that should be *known*. I want to know about how to make that. Now I know how they did it the old way, because I can view in the archives what the early ethnographer Gilbert Wilson recorded when he talked to Wolf Chief about that. So being in these museum spaces is very important – people should have history centres and interpretive centres, like our Interpretive Center that's going up.

My uncle told me there is stuff that gets left out of memory, and that's understandable. He said, "you need to talk to *my* grandpa." That generation made what became Gilbert Wilson's collections. That's the generation I get to see in those archives. So, we go to these spaces with the idea of how our people were, how we used to live and interact, how things were made. These things that were made 120 years ago – that's not that long ago to us. I wish I could have been a fly on the wall in an earth lodge at that time, when they were working on quillwork, making those objects that we used daily. When I see that object in the collection, it brings me to that moment in the lodge. It feels close to me in time, close to those ancestors. Being in the presence of museum collections and archives really matters. Some museum staff understand this – one told the archivist, "he needs to see the originals, not the microfilm." We want to know more about ourselves. I want to ask the museum staff, "Do you know where you come from? Do you understand what this feels like? How hard it was to get here? The respect and spirituality that we have for this stuff?" The label may say B-249, but for me, I've prayed about seeing this old stuff – the photos, the objects. This has affected my life so much.

Royce: One of my first experiences with museums where I started to feel how much was lacking, was going to New York when I was eight or nine. We went to the natural history museum and I saw that we were still grouped with the wildlife section, with the animals. Going in there made me feel so embarrassed, and the dioramas that they had of our people looked really rough. You could tell it wasn't to make us look real keen. I think about us, and we're just really good-looking people! [laughs] But you don't find that in the museums, you can't relate to it. I always thought art galleries were more accessible, and I could relate to the work being created by Native artists, especially at the Institute of American Indian Arts. When I went to the gallery in Santa Fe, seeing the type of work that was being done, I felt prouder there than going into a museum. That's always stuck with me, being so little in New York and seeing that. With the work at the MHA Interpretive Center that's coming up, the question is what type of memories can we make for people? How can we inspire our people? What type of exhibits and what type of materials can we bring into that space, where we can draw out those memories – and make them feel like it's home. I don't want them to feel like our people are a glass exhibit. It should represent something that you feel ownership over, you feel related to. I don't ever want to have that feeling of embarrassment for the representation of our people in a museum. I want the Center to be something really good and positive, where you come out with the feeling like it's home.

I didn't get to grow up past the age of 3 or 4 with my grandmother there. I didn't get to grow up with all of our different type of Hidatsa stories because I didn't live with her anymore. Access is incredibly important, so every one of our tribal members and non-tribal members can experience what we would hope our communities are working towards: healthy relationships, where we're passing that knowledge down to our people and sharing that family. Unfortunately, not all of us have been blessed like that. But we should all have the opportunity, if we want to learn. And, if we want that, there's a place, there's a space to do that.

Mary Baker Price: I don't think that I was personally affected by museums. Why I am at this conference, what drives me to learn about our ways, is that our generation, it seems like it is more important for us to know our identities and who we are, and how we can make ourselves through our societies, or clanships, or whatever. I come from a family that didn't really teach any of that. But I always wanted to know, because I felt a deeper connection with stuff like that. But, as far as [the] Interpretive Center goes, I think that as a people, we need to get away from a lot of different kinds of structures, like nepotism and how politics plays a heavy role in how we carry ourselves or the things that we do. It's important for me to learn the language because I want to be able to speak, I want to be able to teach, I want to be able to pray in the language. I think we come from a generation that was fearless *because* we didn't have to face a lot of things that were illegal. By the time we were born, things were already made legal – we could go to ceremony, we could go to powwows, we could even practice our ways. But, without that knowledge, who we are and where we come from, I think there's a huge generation gap. And I think it's up to us to do all this research, and bring to the table – like, okay, how are we going to find ourselves again? We kind of have to – not necessarily recreate but re-establish ourselves and clanships. I think about our elders and our ancestors and what they had to face was a lot [of] hardship and oppression. That spiritual energy that we come from still reaches out to us, but it has to be reborn out of a place of not having fear, not having traumatic experiences – so that it's out of love. How you carry yourself, it's out of *love* with it. I feel like this conference is a good experience because there's a lot of people that share that same inspiration to repatriate, or bring back the language, or restore that knowledge of medicines and stuff like that. I think we need to be that again.

Royce: The Interpretive Center needs to be that research hub, that teaching hub, that place where our people can come back, and we can learn what we *should* have learned – where we have access to learn the clanships, our family trees, and not just read words on a paper but see those images and hear our ancestors. It *seems* like common sense, like everyone should know how to do just basic things, but we're not taught that. Some of us were not taught that at all. Like, even at the Tribal Education Department, they want to do "Rock Your Mocs," and, you know what? We have tons of families that weren't taught how to make moccasins. Their children don't even have mocs. So, the ones that can't rock their mocs, what are we going to do for them? That's just real simple. There should be classes like that at the Interpretive Center, resources for our tribal members to learn such skills and more, so we can help our people.

Mary: I guess, a question is – what's your vision? What do you want the outcome of the Interpretive Center to be?

Elijah: I want to be in a space where there's people *like me*, there's people that look like me, talk like me, and so forth. People who have more of an understanding that there's an emotional tie to all of this historical stuff – more than, just, we have to keep it temperature controlled, we have to do everything in these guidelines. I *understand* that's an important piece of how to preserve it. What I envision is a space where we can bridge that gap between generations. We can bridge that gap between families that have knowledge of our ways, and our language, and other cultural aspects. I envision kids, youth, coming through, getting inspired. I envision a place that is welcoming and is, like you guys were saying, *home*. I envision a space where we can do workshops. And I feel like it's a perfect opportunity. But I do understand that there's going to be obstacles and other barriers within ourselves, and within these other museums that we're going to be connecting with to bring stuff back. That's just a couple things off the top that I can verbalize right now.

Mary: Maybe we could restore the land to plant some medicinal plants that we work with. Knowing how strong that medicine is, I want us to be able to teach little ones, and bring that plant knowledge back to common knowledge. And it's important to show in the Interpretive Center that this is where we came from. But I don't want it to be limited to just traumatic experiences – this is the Garrison Dam, and smallpox, and stuff like that.

Royce: Yeah, it's important to contextualize it, *now*. Why is it relevant? Why is that knowledge relevant today? My vision, I'm coming at this from that culture and language side. My first language was Hidatsa, that's what was spoken to me and what I remember dreaming in first and speaking first. But that was taken because of intergenerational trauma. And in my story, that's not unique at all. That story is shared by a large number of our people, of the kids in the schools. And I see a common habit of looking back on how good our people had it in the past, and we compare ourselves and our problems and issues of today to how good our ancestors had it. As if that was the best it was and will ever be. Well you know what? We're vibrant *right now*. The people that we work with, who we're inspired by each day, *that's* where it's at. And we have to have a place where we're building ourselves up. Because if I just dwelled on the trauma of my youth, I might not be here. I could be off in some gutter, in jail, or whatever else. But something along the way I experienced inspired me enough to be more, do more. That inspiration and drive to do more is the same with many of our people. So, how can we blend our inspirations, wants, and needs for our future and the vibrancy of our people into a space that provides access to the resources needed for us to continue learning who we are, our culture and language and history? How can we have a space that shows our people are really, *really* great *today*, just as our ancestors were. Our people are doing so much all over our reservation and beyond. Like the Nueta Language Initiative that Elijah worked with – I've been following it on Facebook for a number of years! But because I don't live in Twin Buttes, I didn't know about all the projects happening in the community. I had no idea about the gardens project in Twin Buttes! The gardens, the orchard they're building, the apiary. What fantastic stuff is going on down there, and we can use that model in other areas of our reservation. We've got people right in our own community who have so much knowledge and [so many] skills to share.

But we just have a disconnect with communication. All of the language projects that have been ongoing – there are a lot of other people doing that work. We just have to reinforce and maybe relearn our ancestral ways of collaborating and communicating collectively and cohesively. So, my vision, my goal for the Interpretive Center, is that we at least will have that knowledge in one central space, where we're up-to-date and highlight the amazing work our people are doing. If I want to learn about something – or maybe I don't even know exactly what I want to research – something at the Interpretive Center will intrigue me to find out more. And we'll have access to the people who are teaching it before it's gone, so we can ensure that we have that experience and knowledge. Because there's still lots of really fantastic people in our community that have so much knowledge about cultural and language ways, but somewhere along the way they've either been put off or someone didn't invite them to share their knowledge. But we've got to find a way to mend those relationships.

This is us – 50 more years – I don't know about you [laughs] – women live longer!

Elijah: Talk about negativity! [laughs]

Royce: I mean, we have 50, 60 more years of work ahead of us. And ten years down the line, I don't want to still be talking about *one* day, one day the future generations will have it. I want to experience that in *my* lifetime, to be part of that, to experience it . . . An hour! [We've been] talking for an hour. I think we're a good bunch, man.

Mary: I need to hang out with you more.

Note

1 The conversation was recorded by Jen Shannon. Mike Barthelemy was part of our discussion, and his thoughtful comments helped motivate our discussion. We all have copies of the recording and the original transcript, which was edited and rearranged by Jen for clarity and length down from 25 pages, and then reviewed, edited, and approved by us. Jen began the conversation, then part way through walked away and then we talked amongst ourselves.

6 Reflections on empowering youth in science museums

Marianne Achiam

Recent decades have seen a profound shift in the way museums perceive themselves and how they, in turn, are perceived by their surrounding communities and societies. New technologies, new economic realities, and rapid demographic and generational changes have set the stage for this shift[1] and many museums have responded quickly and decisively by re-appropriating their traditional and self-referential functions to reflect more community-oriented perspectives[2]. However, nineteenth and twentieth century museum logics may still linger in places, leaving members of the non-dominant yet fast-growing communities on the outside.[3]

Museums are not the only institutions undergoing scrutiny. The same winds that buffet what were traditionally the temples of knowledge also act upon that knowledge itself. More and more, assumptions about what constitutes scientific knowledge and the objectivity of that knowledge are being questioned. Haraway offers one of the most compelling critiques of the objective scientific "gaze from nowhere", observing that rather than being objective, this gaze "signifies the unmarked positions of Male and White."[4] Many other scholars have followed in her footsteps, pointing out the gendered, raced, and classed nature of Western science.[5]

Acknowledgement of this social injustice is manifest in the recent and widespread push towards the ideals of Responsible Research and Innovation (RRI) seen in societies and nations across the world, and in particular in the discourse of the European Commission, as outlined in the following essay by Rosenfeld and Blonder. It is at the intersection of these developments that we find the present-day science museums that participated in a European Union (EU)-funded project, called Irresistible, where they opened their doors to science students and offered up their spaces to the exhibits created in the project. The Irresistible Project developed at least two important themes: the implications for science museums exhibiting science produced by non-scientists, and reciprocally, implications for projects that rely on students producing and displaying exhibits in science museums. Small groups of students selected scientific content and engaged in curating exhibits based on questions derived from the EU's Responsible Research and Innovation framework. The resulting exhibits featured socially and historically situated science, dilemmas related to science, and wicked problems with no clear scientific answers, all seen through the eyes of the participating students rather than those of the museum staff. What are the implications of this relinquishment of control for the participating science museums?

The socio-scientific and situated perspective that was prevalent in the student-curated exhibits is not necessarily novel to science museum exhibition designers[6]. However, the surrendering of curatorial authority is probably more unusual. Studies show how participatory practices in museums can be fraught with tension because they are perceived to challenge the expertise of staff members[7]. Further, some science museum professionals see themselves as full members of the scientific community[8]. The adherence to institutionalized notions of science afforded by this membership may further marginalize the roles and priorities of non-scientists in the museum.

Science museums can better serve their publics by adopting the participatory mind-set of Responsible Research and Innovation across the institution and establishing a permanent agenda of co-creating exhibits *with and for* their communities. This agenda would in many cases require deeper and more radical organisational changes to allow for a more equitable sharing of authority.[9] However, such a sharing of authority would not only create stronger ownership of the museum from its community: If it were made explicit, the sharing of curatorial authority would help dispel the "gaze from nowhere" illusion that characterizes much dissemination of science.

Students as curators

The Irresistible Project engaged students in deconstructing the science of scientists and reconstructing it in the form of exhibits intended to engage visitors in critical dialogue about science. In other words, the students who might more often have been on the receiving end of this process of *didactic transposition* (cf. Mortensen[10]) were put into the position of being producers. What are the implications of this role reversal for the students?

Few studies have examined the outcomes of engaging students in curatorial work. One line of inquiry focuses on how the meeting between students and collections-based objects affords scientific discovery processes and how these discovery processes are subsequently used by students as the raw material for the deconstruction and reconstruction of content for an exhibition[11]. Although the students in the Irresistible Project seemed to work with multimodal representations rather than specimens *per se*, a similar observation is hinted at by Kampschulte and Parchmann who write, "Developing an exhibition with students includes a lot of tasks which require the use and application of several of these multimodal representations, while also reflecting on the communicative role of the exhibition.[12]

Taken together, this could mean that when students work directly with scientific objects, specimens, and representations rather than the heavily transposed versions they encounter in educational settings (cf. Clément, Mortensen[13]), new opportunities for scientific inquiry may arise. It is perhaps such opportunities that the Irresistible Project capitalizes on in the meetings between students and cutting-edge science, and that lead to the important learning outcomes.

Another important aspect of engaging students in curatorial work allows me to revisit Haraway's critique of the scientific "gaze from nowhere."[14] Rather than being objective, universal, and constant, scientific knowledge (and values and practices) is always relative to the institution it inhabits.[15] When transposed from one institution to the next – as was the case in the Irresistible Project when students deconstructed the science of scientists and reconstructed it as the science represented in an exhibit – the knowledge in question is transformed and adapted to fit its new institution. This means that when students are engaged in designing exhibits, they are also acquiring the ability to perceive and shift between institutional ecologies. I suggest that this ability effectively provides the involved students with a unique vantage point from which to observe the institutional relativity of scientific knowledge. Certainly, the study by Kreuzer and Dreesmann showed how, as a result of their curatorial activities, students acquired not only scientific content knowledge but also knowledge of the museum institution and its particular *ways* of producing scientific knowledge.[16] And when the students and teachers in the Irresistible Project began to "substitute an exclusively 'objectivist' and impartial view of nature and technology with a more complex and nuanced view," I suspect that part of this outcome can be attributed to a similar flexibility in institutional perspective prompted by the project.

Acknowledgement

The writing of this essay was assisted by a grant from the European Union's Horizon 2020 Framework Programme for Research and Innovation (H2020-GERI-2014–1) under grant agreement No. 665566. The statements made and the views expressed are solely the responsibility of the author.

Notes

1 Black, Graham (2012). *Transforming Museums in the Twenty-First Century*. London: Routledge.

2 Achiam, Marianne, and Sølberg, Jan (2017). Nine meta-functions for science museums and science centres. *Museum Management and Curatorship*, 32(2): 123–143. doi:10.1080/09647775.2016.1266282

3 Dawson, Emily (2014). "Not designed for us": How informal science learning environments socially exclude low-income, minority ethnic groups. *Science Education*, 98(6): 981–1008. doi:10.1002/sce.21133. Robinson, Helena (2017). Is cultural democracy possible in a museum? Critical reflections on Indigenous engagement in the development of the exhibition encounters: Revealing stories of aboriginal and Torres Strait Islander objects from the British museum. *International Journal of Heritage Studies*, 23(9): 860–874. doi:10.1080/13527258. 2017.1300931

4 Haraway, Donna (1988). Situated knowledges: The science question in feminism and the privilege of partial perspective. *Feminist Studies*, 14(3): 575–599. doi:10.2307/3178066

5 Brickhouse, Nancy W. (2001). Embodying science: A feminist perspective on learning. *Journal of Research in Science Teaching*, 37(5): 441–458. Wong, Billy (2016). *Science Education, Career Aspirations and Minority Ethnic Students*. Hampshire: Palgrave Macmillan.

6 Achiam, Marianne, and Sølberg, Jan (2017). Nine meta-functions for science museums and science centres. *Museum Management and Curatorship*, 32(2): 123–143. doi:10.1080/09647775.2016.1266282

7 Bønnelycke, Julie, Sandholdt, Catharina Thiel, and Jespersen, Astrid Pernille (2018). Co-designing health promotion at a science centre: Distributing expertise and granting modes of participation. *CoDesign*. doi:10.1080/15710882. 2018.1434547

8 Feinstein, Noah Weeth, and Meshoulam, David (2014). Science for what public? Addressing equity in American science museums and science centers. *Journal of Research in Science Teaching*, 51(3): 368–394. doi:10.1002/tea.21130

9 Kinsley, Rose Paquet (2016). Inclusion in museums: A matter of social justice. *Museum Management and Curatorship*, 31(5): 474–490. doi:10.1080/09647 775.2016.1211960

10 Mortensen, Marianne Foss (2010). Museographic transposition: The development of a museum exhibit on animal adaptations to darkness. *Éducation & Didactique*, 4(1): 119–137.

11 Powers, Karen E., Prather, L. Alan, Cook, Joseph A., Woolley, James, Bart Jr., Henry L., Monfils, Anna K., and Sierwald, Petra (2014). Revolutionizing the use of natural history collections in education. *Science Education Review*, 13(2): 24–33. See also Achiam, Marianne, Simony, Leonora, and Lindow, Bent Erik Kramer (2016). Objects prompt authentic scientific activities among learners in a museum programme. *International Journal of Science Education*, 38(6): 1012–1035. doi:10.1080/09500693.2016.1178869. Kreuzer, Pia, and Dreesmann, Daniel (2016). Museum behind the scenes – an inquiry-based learning unit with biological collections in the classroom. *Journal of Biological Education*, 1–12. doi:10.1080/00219266.2016.1217906

12 Kampschulte, Lorenz, and Parchmann, Ilka (2015). The student-curated exhibition – a new approach to getting in touch with science. *LUMAT*, 3(4): 462–482, 464.

13 Clément, Pierre (2007). Introducing the cell concept with both animal and plant cells: A historical and didactic approach. *Science & Education*, 16: 423–440. Mortensen, Marianne Foss (2010). Museographic transposition: The

development of a museum exhibit on animal adaptations to darkness. *Éducation & Didactique*, 4(1): 119–137.

14 Haraway, Donna (1988). Situated knowledges: The science question in feminism and the privilege of partial perspective. *Feminist Studies*, 14(3): 575–599. doi:10.2307/3178066

15 Chevallard, Yves, and Bosch, Marianna (2014). Didactic transposition in mathematics education. In *Encyclopedia of Mathematics Education*, edited by S. Lerman, 170–174. Dordrecht, Netherlands: Springer.

16 Kreuzer, Pia, and Dreesmann, Daniel (2016). Museum behind the scenes – an inquiry-based learning unit with biological collections in the classroom. *Journal of Biological Education*, 1–12. doi:10.1080/00219266.2016.1217906

7 Promoting responsible citizenship in science museums through student curated exhibits

Sherman Rosenfeld and Ron Blonder

Since the last decade of the twentieth century, the digital revolution and changes in perceptions of science and society have presented science museums with several challenges. The far-reaching effects of the digital revolution are based in large part on digital technology's increased storage of information and how it creates greater interconnectedness and faster communication between people. Social media tools have given people around the world the ability to co-create shared content. These widely available tools have essentially turned consumers into producers and passive recipients of knowledge into knowledge creators. This development presents science museums with the challenge of how to engage and empower their visitors, who are becoming more and more literate with these creative and shared tools.

At the same time, perceptions of science and society have been evolving. A growing consensus views science as (1) an integrated STEAM approach, in which the sciences, technology, engineering, art/design, and math are understood as interrelated disciplines, and (2) a process based on diverse interests and values that are connected to overlapping societal, economic, political, and ethical issues. These developments present science museums with the challenge of presenting an integrated STEAM approach that includes socio-scientific issues to the public.

When the digital revolution and the changes in perceptions of science and society are taken together, the guiding question becomes: How might science museums engage their visitors as creators of exhibit content that is integrated with socio-scientific and often controversial issues? One promising approach to this question is to engage students to create exhibit content, based on cutting-edge scientific topics and a new framework for a science–society partnership, as defined by the European Union (EU).

The RRI framework and student curated exhibits

Responsible Research and Innovation (RRI) is a framework developed by the EU to: (a) guide diverse stakeholders – science, technology, and the general public – to productively interact with each other to advance scientific research and technological innovation, and (b) enhance the public's level of trust and participation in this effort. The RRI framework evolved in the early twenty-first century, largely as the result of two developments: (1) the emergence of new cutting-edge technologies (e.g., genetically modified organisms and synthetic biology, information and communication technology, robotics, geoengineering, and nanotechnology), and (2) the motivation of innovation policy actors to develop these technologies for social benefit and to address public anxiety over their unintended and irreversible consequences.[1] RRI was adopted as the key action of the "Science with and for Society" objective within Horizon 2020, an EU Research and Innovation programme with nearly €80 billion of funding over seven years spanning 2014 to 2020.

One way to connect the RRI framework to students and other societal actors is through student curated exhibits. In the EU-funded Irresistible Project, ten European partners joined forces to answer a call from Horizon 2020, in order to develop an innovative approach to introduce students to cutting-edge research and development efforts and their influence on modern society, via the RRI framework. Each of the ten partners designed and prepared a curriculum module in which the students explored cutting-edge science and the six RRI dimensions by using an inquiry teaching model. Each module was developed by a Community of Learners (CoL) composed of a research scientist, high-school science teachers, a member of the local science centre, and science educators. Each module was based on the research work of a research scientist at the university and reflected a local expertise of the researchers in the participating university. In each case, the students presented their knowledge to other students and members of the general public through student-curated exhibits that were displayed in their schools and in local science museums. The ten modules were pilot-tested twice, revised, and translated into English; they appear on the project website.[2]

For example, a lesson called "The Story of Lead" was used to introduce Israeli students to technological developments (e.g., leaded wine, leaded paint, and leaded gas) that led to negative and unintended consequences, largely because the dangers of these developments occurred without attention to the RRI dimensions of engagement, open access, ethics, science education, gender equality, and governance.[3] This lesson was a catalyst for the students and their teachers to deepen their understanding of the six RRI dimensions by asking questions about them (Table 7.1).

Each RRI-based curriculum in the Irresistible Project used a modified inquiry teaching model, based on the 5E model.[4] This model was chosen because it provides a good structure for inquiry-based curriculum

Table 7.1 The six dimensions of the Responsible Research and Innovation (RRI) framework and their interpretations by students.

RRI framework dimensions	Student questions
Engagement. All societal actors (researchers, industry, policy makers, and civil society, including students and teachers) should responsibly participate in scientific research and technological innovations.	Which public organizations should be involved in the research or innovation? Are the voices of everyone involved equal in the decision-making process? What is the decision-making process? To what extent (if at all) should people who are not knowledgeable about science influence scientific decisions?
Open Access. Results of publicly funded research and development should be made available openly and freely (e.g., via free online access) to increase the use of scientific results and the exposure to technological developments.	Is it enough to publish research results in professional journals that are accessible only to the scientific community? If not, how might research results be presented so that they are available to the general public? Should studies also publish possible shortcomings and risks of the innovation? Should there be an obligation to publish information about patents?
Ethics. All aspects of research and innovation should be conducted according to ethical norms in order to increase their societal relevance and the acceptability of these outcomes.	Which ethical values are essential to consider? Does adhering to ethical standards improve research or hinder it? To what extent does the product and its development take into account social and environmental values? Is the development sustainable? Does it take into account possible effects on the future?
Science Education. Students and the general public should be able to learn about the processes and outcomes of research and innovation efforts.	What degree of commitment (if any) should the scientist have to science education? How much effort should scientists and technologists be asked to invest in order to share their research and development with people who are not experts in these areas?
Gender Equality. Gender balance in research teams should be established, in order to close the gaps in the participation of women.	What is the proper representation of men and women in research and development? What should happen if there is no proper representation of men and women?
Governance. Research and innovation practices should be overseen by governmental organizations that integrate the above five dimensions in their decision-making process.	Who will supervise the work? What stages of research and development need to involve supervision? What should be the source of authority for this supervision? Do scientists and technologists have an obligation to report their work? What specifically should be involved in the process of supervision?

Information in Table 7.1 is based on Blonder et al. (2016).

development and teaching as well as a motivating invitation for student creative work. The model structures the curriculum according to the stages of *Engage* (providing the trigger for student involvement), *Explore* (providing opportunities for student exploration), *Explain* (presenting the relevant concepts), *Elaborate* (applying the concepts to new situations), and *Evaluate* (evaluating student learning). The modification was an additional "E" that was added to this teaching model to represent an *Exchange* phase, in which students exchanged their opinions with each other, regarding one or more of the previously mentioned RRI-related questions. During this additional stage, the students designed and constructed exhibits based on topics they chose. In doing so, they invited the exhibit visitors to learn about the scientific aspects of the topic, as to their understanding of the reciprocal relationship between this science and the society in which this knowledge developed.

For each country, a cutting-edge area of research and development at the participating institution was chosen, since one goal of the Irresistible Project was to expose students to new frontiers of science and technology. The topics of the student-curated exhibitions were then chosen and developed by small groups of participating students and later displayed by the participating schools and science museums. For example, in the Netherlands, the scientific research topic chosen was the role of oligo-saccharides in mother's milk. The research showed that the majority of oligo-saccharides cannot be digested by human babies. This paradox leads to the conclusion that natural human oligo-saccharides are needed for the development of certain beneficial microflora in the infant intestines. This discovery was followed by other studies that developed procedures to synthesize these oligo-saccharides in order to use them as additives to baby formulas. Students designed exhibits that revealed the different ingredients of milk from different sources, exhibits about the bacteria in the intestinal tract, and exhibits about the process of manufacturing baby formula. In addition to presenting this scientific and technological knowledge in their exhibits, students asked the visitors to think about possible dilemmas, such as the possible contribution of the new baby formulas to the independence of women, who as a result can return earlier from maternity leave when the commercial milk formula is similar to mothers' milk. Other dilemmas that were presented in the exhibits related to the approval to sell the milk formula to foreign workers (who sent it back home), which led to a lack of the new formula in the Netherlands.[5]

Another example of the integration of student curated exhibits and RRI was a curriculum unit developed in Israel about an innovative solar cell based on perovskite, that is used to increase energy efficiency. The students were invited to decide whether to replace the windows in their school with these new solar cells. While they explored the topic, they found out that the perovskite-based solar cells include lead, a highly toxic element. Many questions were generated such as how much lead is considered a toxic dose and the economic pricing of these solar cells. The students built exhibits that exposed visitors to the historical story of lead from the Roman days until the present, along with its poisonous effects. They also connected the discussion about lead in the perovskite-based solar cells to journalistic research that was published in Israel about the presence of lead in coffee machines. The visitors to these exhibits were invited to explore the effects of lead on people throughout history. The visitors were also asked to explore the guiding question: "Under what conditions, if any, would you agree to replace the windows of your school with perovskite-based solar windows?"[6]

The topic of alternative energy sources and photovoltaic cells was also chosen as the basis for student-curated exhibits in Italy. Students investigated problems caused by the massive use of fossil fuels (e.g., the environmental issue of global warming, the political issue of being dependent on the oil-supplier countries) and alternative energy resources, (e.g., building a solar panel to produce electricity from sunlight). The resulting exhibits included games that engaged the visitor in different dilemmas based on questions related to the six RRI dimensions. The visitors were invited to play the games that on the one hand provided basic scientific information, and on the other hand, asked the players to consider RRI questions that are related to the possible use of solar energy. Another environmental topic was plastic waste in the ocean, chosen by the German team. Students researched the huge amount of plastic waste in the oceans, its non-biodegradable nature, and its influence on the environment. The inquiry activity included an out-of-school visit to the seashore and collecting the plastics found on the beach. At school, the students analysed the plastics and realized that the majority of what they found will stay stable for hundreds of years and therefore have a severe influence on the ocean flora and fauna. Their exhibit was constructed from the plastic waste that they had collected on a local seashore, which showed its negative influence on marine animals.

The underlying assumption of the Irresistible Project, and similar RRI-based educational efforts, is that if scientific and technological efforts are developed within the framework of RRI, then the resulting awareness and practice can lead to fewer catastrophes caused by the applications of science and technology as well as to more participation and "ownership" by diverse societal groups in these fields. How did the experience of engaging in RRI and student curated exhibits actually affect the attitudes of the participating students and teachers? We addressed this question through the development, validation, and use of a questionnaire for evaluating student and teacher attitudes. In a pre–post study, we found that the experience of engaging in student-curated exhibits resulted in statistically significant differences in positive attitudes regarding all six RRI dimensions described in Table 7.1 for both students and teachers in the participating ten countries. In other words, students and teachers began to substitute an "objectivist" and impartial view of the nature of science and technology with a more complex and nuanced view that incorporates socio-political aspects.[7]

Responsible science citizenship in the science museum

One of the central goals of education – to prepare students to be informed, responsible, and active citizens – should apply to the fields of science and technology. Science education needs to expose teachers and their students not only to the facts, principles, and discoveries of science and technology, but also to their socio-economic contexts, which can have significant social consequences. For example, students should learn that science and technology produce consumer products that are often developed by commercial sponsors of research who have vested interests in these products. Students should learn about this phenomenon and learn to ask such questions as: To what extent can I trust the claims written on the boxes of my favourite breakfast cereals? Are these claims supported by solid scientific research? How can I find out?

The Irresistible Project has shown that students can be guided to practice science citizenship by being given opportunities to (1) investigate the scientific and technology aspects of socio-scientific issues from a wide variety of information sources; (2) care about these issues and the people impacted by them, while understanding the views, needs, and interests of diverse stakeholders; (3) identify and manage the powerful emotions often generated by these issues; and (4) take responsible action and evaluate its effects.[8] This project also demonstrates that integrating social media tools into the design of student-curated exhibitions can amplify both the students' learning and their effectiveness in dealing with these issues. We suggest that the treatment of complex and often controversial socio-scientific issues, using the RRI framework to create student-led exhibitions, can advance responsible science citizenship. A strong partnership between schools and science museums can promote this approach.

To what extent can and should science museums actively promote this agenda? Science museums can better serve their visitors by promoting this agenda in their exhibits and public programs. By doing so, they will become more attractive and relevant to their visitors. At the same time, they will adopt a more contemporary view that sees the practice of science and technology as a complex enterprise that consists not only of facts, principles, and phenomena, but also one that involves overlapping societal, economic, political, and ethical issues.

The integration of RRI and student-curated exhibits can be used by science museums to help visitors think critically about the socio-political aspects of science and technology, in order to promote science citizenship. Ideally, these potentially challenging topics should be a regular feature of science museum exhibitions. A more conservative approach would be to create "public voice" spaces in science museums where exhibits dealing with controversial issues can be displayed. These spaces in museums can be analogous to opinion pages in newspapers. Just as opinion pages host writings that deal with controversial issues, not identified with the newspaper itself, so might these public voice spaces contain exhibitions on controversial socio-scientific issues that present positions not necessarily identified with the museum.

Student-created exhibitions about the scientific and RRI aspects of cutting-edge science research can catalyse new science museum exhibits. Museum staff can develop outreach programs with students and teachers in schools to produce exhibits. This co-design process could lead to the development of exhibitions presented in the schools and the science museums.

In conclusion, we began this essay by posing a guiding question, based on the digital revolution and changes in perceptions of science and society: How might science museums engage their visitors as creators and providers of exhibit content dealing with socio-scientific and often controversial issues? Our suggestion is to combine the RRI framework and student-curated exhibitions as a practical framework to deal with current socio-scientific issues that are relevant to the science museum visitors. What conditions in schools and museums will best support the implementation of this suggestion? What corresponding challenges need to be overcome and what are possible solutions? We leave these questions open for discussion.

Notes

1 Sutcliffe, Hilary (2011). *A Report on Responsible Research and Innovation for the European Commission*, available at http://ec.europa.eu/research/science-soci ety/document_library/pdf_06/rri-report-hilary-sutcliffe_en.pdf. Guston, David H., Fisher, Erik, Grunwald, Armin, Owen, Richard, Swierstra, Tsjalling, and van der Burg, Simone (2014). Responsible innovation: Motivations for a new journal. *Journal of Responsible Innovation*, 1(1): 1–8.
2 From the Irresistible Project website: Engaging the Young with Responsible Research and Innovation. http://www.irresistible-project.eu/index.php/en/resources/teaching-modules.
3 Blonder, Ron, Zemler, Esty, and Rosenfeld, Sherman (2016). The story of lead: A context for learning about responsible research and innovation (RRI) in the chemistry classroom. *Chemistry Education Research & Practice*, 17: 1145–1155. doi:10.1039/C6RP00177G.

4 Bybee, Rodger W., Taylor, Joseph A., Gardner, April, Van Scotter, Pamela, Powell, Carlson J., Westbrook, Anne, and Landes, Nancy (2006). *The BSCS 5E Instructional Model: Origins and Effectiveness*. Colorado Springs, CO: BSCS.

5 Apotheker, Jan, Blonder, Ron, Akaygun, Sevil, Reis, Pedro, Kampschulte, Lorenz, and Laherto, Antti (2017). Responsible research and innovation in secondary school science classrooms: Experiences from the project irresistible. *Pure and Applied Chemistry*, 89: 211–219.

6 Blonder, Ron, Rosenfeld, Sherman, Rap, S., Apotheker, Jan, Akaygün, Sevil, Reis, Pedro, Kampschulte, Lorenz, and Laherto, Antti (2017). Introducing responsible research and innovation (RRI) into the secondary school chemistry classroom: The irresistible project. *Daruna*, 44: 36–43.

7 Blonder, Ron, Rap, Shelley, Zemler, Esty, and Rosenfeld, Sherman (2017). Assessing attitudes about responsible research and innovation (RRI): The development and use of a questionnaire. *Sisyphus: Journal of Education*, 5(3): 122–156.

8 Hodson, Derek (2014). Becoming part of the solution: Learning about activism, learning through activism, learning from activism. In *Activist Science and Technology Education: Cultural Studies of Science Education*, edited by Larry Bencze and Steve Alsop, vol. 9. Dordrecht: Springer.

8 Conflicts between traditional knowledge systems and Western science

Do we need to revise our thinking in order to engage youth from all heritages?

Laura Huerta Migus

There have been as many different systems of thought and knowledge about the natural world as there are human cultures and societies. Indeed, the current definitions of scientific thought and science itself, whose historical roots are 500 or so years old, are quite new when measured against the scope of human history. The cultural roots of the modern scientific narrative are also dominated by the development of empirical methods in Western Europe during the two-century period of the Scientific Revolution (late 1500s through late 1700s). Who is widely considered to be the father of modern mathematics? Newton. Who are the most famous early astronomers most people know? Likely, Galileo, Copernicus, or Halley. It is also no coincidence that the period of birth of modern science aligns with the era of global trade and colonization focused on the lands of North America, which brought Europeans in contact with indigenous flora, fauna, materials, and cultures that were all seen as obstacles to eradicate and fodder to drive European economies. Dominant discourse on science history and scientific knowledge rarely draws these connections, except to refer to the artefacts of North America as exotic materials to advance theories of nature and validate (or dismiss) the claims of meaning by Indigenous peoples. The mindset still exists today that Indigenous knowledge of the natural world is not significant until validated through European scientific empirical methods. These biases are also pervasive in current scientific practice and in both formal and informal science education. Why, for example, are we so familiar with the names of European astronomers, and we know so little of the foremost Mayan astronomers? Is it common knowledge that Pacific Islanders were some of the most sophisticated navigators in the world? These are just two small examples of how the Eurocentric narrative of science has made Indigenous scientific knowledge invisible.

In the last twenty years, Indigenous peoples around the world, bolstered by the 2008 United Nations Declaration on the Rights of Indigenous Peoples, are actively working with such "Western" methods as intellectual property law to undo this exploitation and to protect their traditional knowledge and practices.[1] This will require explicit acknowledgement that the dominant scientific narrative is one that solely emphasizes Western culture and wilfully ignores the implicit cultural imperialism. This recognition is critical for authentic engagement on the issue of multiculturalism in science centre practice.[2] For this reason, the remainder of this chapter will refer to what is commonly called "science" as Western Science, and Indigenous peoples' ways of understanding the natural world as Indigenous Knowledge Systems.

What role do science centres play?

It is especially important to recognize the relative newness of modern science when seeking to assess how science centres and museums communicate science to the public. Science centres and museums are modern institutions with current practices centered on communicating scientific concepts through interactive exhibits and hands-on programming. They focus on phenomenological learning or learning by experiencing rather than a didactic approach.[3] It follows that most science centres have inspired visitors to create and/or nurture positive relationships with Western Science as core to their mission and practice. This frame implies suspect assumptions regarding the initial state of science centre audiences: visitors are seen as either starting from a neutral or a positive relationship to Western Science with a seeming potential to grow from their museum experience. These assumptions often ignore *why* public audiences may have no relationship to Western Science. They also do not acknowledge that many visitors begin with a negative relationship with Western Science, one rooted in the historical realities of the appropriation, violence, and oppression inflicted on Indigenous peoples and other minority groups.[4] It is this author's opinion that this near-total focus on communicating only the positives of Western Science puts science centres at a disadvantage in their efforts to better serve and engage individuals from marginalized communities.

It is reasonable to think that addressing these inequities requires science centre staff to have different capabilities and use different strategies than they would when engaging non-minority audiences. In fact, one should question whether creating a positive relationship with Western Science is even an appropriate goal for science centre outreach. Would shifting the content of science centre exhibits and programming to better reflect the diverse roots of scientific thought across human diversity be more appropriate for all audiences?[5] This chapter explores strategies that range from those sympathetic to Western Science to those that embrace the science centre as a platform for sharing the variety of human scientific practice. Examples are generally drawn from North American science centre efforts to integrate Indigenous Knowledge Systems, given the large number of science centres in North America in comparison to the rest of the world.

Strategies for equity and relevance

Despite the inherent challenges, there have been a number of efforts by science centres to broaden the narrative of science to include non-Western scientific legacies and achievements. Such efforts generally take the following forms: specialized programs for Indigenous youth; efforts that seek to create change on an individual level, primarily structured as professional development programs for science centre staff; exhibits and programs developed for the general public highlighting Indigenous Knowledge Systems; and organizational approaches in which the science centre explicitly integrates both Indigenous Knowledge Systems and Western Science perspectives throughout programming and exhibits.

There are also important exemplars of bringing together Indigenous Knowledge Systems and Western Science outside of the science centre setting. Led by Indigenous peoples, these provide inspiration for how authentic and equitable approaches to incorporating Indigenous Knowledge Systems should be developed. As such, these efforts disrupt historical oppressions and successfully engage Indigenous youth positively in science learning and discourse.

Youth programs for Indigenous youth

It may seem obvious that programming geared toward direct engagement of Indigenous youth in positive science experiences should be the first option for new and expanded programming in science centres. There are, however, few exemplars, and those vary in their design, desired impacts, and sustainability.

Salmon Camp. The Oregon Museum of Science and Industry (OMSI) and the Native American Youth and Family Association (NAYA) partnered from 2003 to 2011 to offer the *Salmon Camp Research Team*.[6] This program focused on introducing middle- and high-school-aged Native youth to salmon habitat recovery and restoration practices from both Indigenous Knowledge Systems and Western Science. The program was structured around residential field research and, in later years, afterschool extensions focused on community and family engagement. This program model is one of the most intensive programs for engaging youth offered by any science centre. For more than eight years, OMSI and NAYA offered week-long research experiences for groups of less than 20 Native American youth, many of whom participated in *Salmon Camp* multiple times. Development and implementation of this program, which was funded externally, required an incredible investment of resources so that staff from NAYA and OMSI could build the mutual trust and understanding necessary to develop a curriculum that seamlessly integrated Indigenous Knowledge Systems and Western Science. It also needed to overcome significant structural obstacles to youth participation, even though the program was free to participants. This effort gave participating Native youth the opportunity to learn about cutting-edge technologies used in natural resource management practices (GIS, GPS, research databases) alongside traditional approaches to habitat management, centering on the importance of salmon as a core natural resource for the various Indigenous cultures of the region. Youth were introduced to these technologies by tribal elders/knowledge holders and research scientists from both cultural paradigms. Through these experiences, youth and families reported positive outcomes in all areas: increased knowledge and value of Indigenous Knowledge Systems, increased sense of self-efficacy with respect to Western Science, and positive attitudes toward pursuing post-secondary education.

Native Youth in Science – Preserving Our Homelands. A collaboration was launched in 2012 between the Mashpee Wampanoag Tribe (Massachusetts), the Woods Hole Coastal and Marine Science Center (WHCMSC), and the US Geological Survey (USGS) to develop a six-unit, six-week summer program for middle school youth from the Mashpee Wampanoag tribe. This program was structured around a "home base" for Native youth, where the first day of each unit was opened at a tribal office and led by tribal leaders, followed by field trips to various USGS and WHCMSC sites. Each unit of the program was structured to weave together critical cultural Wampanoag narratives (tribal creation story, role of water, oral histories of settlement) with Western Science concepts (water quality, geology, topography). Project leaders explicitly acknowledged the importance of grounding all Western Science content within the acceptable Wampanoag cultural protocols of knowledge sharing and as support for the validity and value of Indigenous Knowledge Systems. In support of these parameters, scientists at the various research sites tapped to contribute content to the program were provided with training by project team members on cultural protocols to support mutual respect of both ways of knowing. Further, lunch each day was provided by tribal members and usually consisted of traditionally sourced and prepared food, which served to ground the experience in the cultural history and practices of the Wampanoag and their relationship to the natural world. This program, while offered for multiple years, was also resource intensive: more than 20 tribal representatives and USGS and WHCMSC staff were involved in the initial effort which served 14 youth.

These two exemplars of direct service to Native youth offered by science centres, while limited in their scope, offer important points for consideration. They emphasized an equal partnership between the Native community and the science centres. In both previous examples, project teams were explicitly described as balanced and equal partnerships between Indigenous and Western knowledge holders and project representatives. This type of relationship, built on long-term trust and mutual understanding, is necessary for the development and implementation of programming that creates a narrative of the equal value of Indigenous Knowledge Systems and Western Science practices.

Indigenous Knowledge Systems are grounded in Indigenous peoples' deep historical and spiritual connection to their geographical homelands, and has been traditionally handed down through real-life (as opposed to classroom) learning experiences. Both programs summarized here incorporate hands-on, immersive field experiences to engage youth in both Western Science and Indigenous Knowledge Systems learning.

In both of these programs, consideration is given to the cultural protocols of both tribal communities and Western Science (scientific method, data collection protocols, etc.), with adherence to tribal cultural protocols as the primary filter. This feature is perhaps the best reflection of how science centres can acknowledge the historical oppression of Indigenous peoples. Incorporating science content as complimentary to Indigenous Knowledge Systems is a critical strategy for healing the generational trauma inflicted on Indigenous peoples through Western colonization.

Unfortunately, although both of these direct service programs reflect best practice in engagement and show positive impact on both the professional staff and youth involved, it is not evident that either has effected substantive change in the way that science centres present content in such a way as to incorporate Indigenous Knowledge Systems.

Challenging professional assumptions

The efforts discussed in this section target the staff from science centres and other science-focused institutions (planetariums, astronomical societies, universities, etc.) grounded in Western Science paradigms. They provide opportunities for these individuals to learn from the Indigenous cultures who hold knowledge around bridging Indigenous Knowledge Systems and Western Science paradigms.

1   **Cosmic Serpent.** The *Cosmic Serpent* project, launched in 2008, was a collaboration between the Indigenous Education Institute, the

University of California, Berkeley Space Sciences Institute, the Association of Science-Technology Centers, the Institute for Learning Innovation, and Native Pathways.[7] It created a series of geographically organized professional development experiences for science centre professionals and tribal educators to explore how to bridge Indigenous Knowledge Systems and Western Science paradigms in informal, non-classroom settings. Over the four years of the project, more than 100 individuals from tribal communities and science centres participated in a series of two week-long regional workshops in California, the Southwest, and the Pacific Northwest. These workshops explored the relationships between Indigenous Knowledge Systems and Western Science from both a historical perspective (the impact of Western colonization and science on Indigenous peoples) and a content perspective (different kinds of knowledge on topics such as astronomy, water, botany, etc.). Each regional workshop was intentionally organized to elevate Indigenous Knowledge Systems paradigms in science learning. They culminated in a final convening that brought together representatives from all three regions.

Participants were hosted by local tribes and experienced local Indigenous scientific knowledge through immersive experiences that followed appropriate cultural protocols. A stand-out feature of this program was the real-life experience of non-Indigenous participants functioning in a professional development environment that operated according to Indigenous cultural practices. For example, in most typical Western professional development experiences, participants are introduced to new program models and content and are expected to take these learnings back to their organizational environments to share and implement. The expectation to directly "port" knowledge from one environment to another is a primary factor in determining the value of many typical professional development experiences. In contrast, native knowledge holders that shared practices in the *Cosmic Serpent* workshops were explicit that the presentations were proprietary cultural knowledge and could not be taken back and incorporated into science centre programming without further discussion and permission. This new orientation to knowledge sharing was found by non-Indigenous participants to be the most important and most challenging learning of their experience.

The assumption by science centre staff that knowledge could be taken and re-shared outside of the context of the workshops illustrates an important cultural difference between Western Science and Indigenous Knowledge Systems. In the context of Western Science, there is the expectation that all scientific knowledge and learnings should be shared publicly, under the assumption that free access to all knowledge is best for society. In contrast, Indigenous Knowledge Systems consider that the transmission of knowledge is governed by specific protocols and cultural practices in order to ensure that such knowledge is understood and used in appropriate ways to best serve the survival of the community and the treatment of the natural environment. For example, certain types of stories or kinds of knowledge are only taught at specific times of the year and to particular people. This contrasts the "transactional approach" of Western Science with a more "relational approach" characteristic of Indigenous Knowledge Systems. Participants worked to bridge these differences in cultural practice by leveraging the personal relationships developed through the project, and these in turn led to institutional relationships that enabled adherence to and respect for cultural protocols around Indigenous Knowledge Systems at a systemic level.

2   **Native Universe.** Building on the learnings from the *Cosmic Serpent* project, the same collaborative project team launched the *Native Universe* project[8]. This offered three, nine-month residencies at three science centres/museums in an effort to increase their capacity to communicate to the public about environmental change and human relationships to the natural world from both Western Science and Indigenous Knowledge Systems perspectives. The project wanted to transition from *individual* capacity building achieved in the *Cosmic Serpent* project to *organizational* capacity building. The participating organizations all had staff that had been a part of the *Cosmic Serpent* project. Through a similar approach, each science centre/museum engaged their entire staff in a series of professional development experiences that explored the history and experiences of local Indigenous people with respect to Indigenous Knowledge Systems and their relationship to Western Science. These experiences resulted in authentic relationship building between science centre/museum staff and local Indigenous communities. At the end of each residency, participating museums reported positive gains across all of these trajectories.

Examples of professional development interventions for increasing the capacity of science centres to integrate Indigenous Knowledge Systems into their offerings are few and far between. It is notable that these two efforts are led by the same institutional partners and have been implemented only recently. Similar to the youth programs outlined earlier, both of these programs were grounded in principles of equalizing the relationship between Indigenous Knowledge Systems and Western Science. They involved field experiences, and they struggled with the tensions between Western Science and Indigenous Knowledge Systems practices around knowledge sharing and translation. These interventions also required significant resource investment from external sources to support the intensive relationship building between individuals and organizations. In both cases, participants indicated positive and sustainable *individual change* with respect to the capacity for integrating Indigenous Knowledge Systems and Western Science content and relationship building, but they also expressed concerns about the ability of their home institutions to sustain such engagement financially and in the face of personnel changes.

Creating content for the public

By far the most widespread strategy for acknowledging and integrating Indigenous Knowledge Systems into the science centre environment is through permanent or traveling exhibits. This is arguably the most effective strategy for acknowledging the value and contributions of Indigenous Knowledge Systems to humanity's collective scientific knowledge. However, many science centre approaches still ghettoize Indigenous Knowledge Systems as subservient to Western Science content, rather than as complementary to it.[9] Although a number of science centres – especially those located in geographies with significant historical and demographic presence of Indigenous peoples – have permanent exhibitions referring to practices of local Indigenous peoples, these exhibitions are often installed in separate, stand-alone galleries and are not referenced in other areas of the museum. And, although most of these exhibitions are curated along best practices through the paradigm of Indigenous partners and according to cultural protocols, they often position Indigenous Knowledge Systems and Indigenous peoples as artefacts of the historic past, not as current contributors to humanity's collective scientific knowledge.[10] If and when individuals from Indigenous communities are highlighted, it is often those, such as research scientists, who participate in the Western Science enterprise. This has the effect of extending a narrative that mostly upholds the dominance of the Western Science paradigm: see, Indigenous people can be scientists too! Although these exhibitions may have limited value, they do raise awareness of the complex and rich roots of Indigenous Knowledge Systems, and they indicate the long-term commitments of museums to inclusion and respect of local Indigenous communities. It is critical to understand, however, that non-Indigenous visitors are the primary intended audience for these exhibits. Although there may be specialized programs, such as facilitated field trips and community-centered events, that are offered to Indigenous audiences, it is not clear how effectively they nurture positive outcomes for Native youth.

Traveling exhibitions and programs are another strategy for incorporating Indigenous Knowledge Systems into science centre content. Traveling exhibitions can be an effective way for science centres, whose content is dominated by the Western Science narrative, to incorporate Indigenous Knowledge Systems in purposeful, if limited, ways. The best of these exhibitions include *The Roots of Wisdom*, developed by OMSI, IEI, the Tulalip Tribes, Revitalization of Traditional Cherokee Artisan Resources, Portland Art Museum, the Pacific American Foundation, and the Confederated Tribes of the Umatilla Indian Reservation. Another example is *Yuungnaqpiallerput (The Way We Genuinely Live): Masterworks of Yup'ik Science and Survival*, developed by the Anchorage Museum in partnership with the Calista Elders Council and the Smithsonian Institution.[11] In the case of both of these exhibitions, the developers intentionally included programming, such as specialized curricula and outreach guidance, and they prioritized installations at tribal museums and cultural centres, both of which helped to positively engage youth from Indigenous cultures in science. Unfortunately, these exhibitions are primarily touring to science centres and museums, whose regular audience is mostly from non-Indigenous backgrounds, and thus their impact on engagement with Native youth is likely low.

Planetarium shows are one programmatic effort by science centres that does show promise in mediating the relationship between Indigenous Knowledge Systems and Western Science. Indigenous peoples all over the world have incorporated into cultural stories, spirituality, and environmental practices the deep and complex cosmologies that align with highly localized interpretations of astronomical phenomena.[12] Planetarium shows are arguably the closest experience to a traditional storytelling experience offered in most science centres: the audience is gathered to watch and listen to a single narrator and in a visually immersive environment. They offer qualitatively different types of learning experiences than either programs or exhibitions in science centre environments. Planetariums are very popular, engaging visitors of all ages, and a menu of shows is offered in a single day, which allows for integration of a variety of narratives with relative ease. In addition, star stories are a universal cultural meme, from the Greek myths that give Western constellations their names to the sacred stories of the Navajo and their constellations that mark the story of creation and the changing of the seasons. Another advantage of planetarium shows is that, because they are media products, there is no need to worry about the sourcing, protection, and use of cultural objects. There does remain, however, the challenge of honouring the cultural protocols and the intellectual property rights of Indigenous Knowledge Systems. This may mean that certain shows are only offered during the time(s) of the year when they are traditionally shared in the originating culture, or at a certain time of day. Finally, when it comes to engaging Native youth in learning experiences that integrate Indigenous Knowledge Systems and Western Science, planetarium shows have the advantage of being quite mobile. Many science centres and local astronomical societies use portable digital dome theatres, which travel to community sites, such as schools, civic halls, and community centres, at relatively low cost. This allows for this content to come to youth in their own communities, mediated by their own elders, and at times that align with appropriate cultural celebrations and observances.

Ideally, every science centre would explicitly incorporate multiple cultural legacies and worldviews of human scientific thought throughout their offerings. In fact, there are just a few that have taken this approach, for example, the Museum of New Zealand/Te Papa Tongarewa, the 'Imiloa Astronomy Center of Hawaii, and the Bernice Pauahi Bishop Museum.[13] The details of how each institution came to integrate in their mission both Indigenous Knowledge Systems and Western Science are worth knowing. All three of these institutions educate non-Indigenous publics about the cultural richness and complex scientific knowledge of local Indigenous Knowledge Systems and serve as cultural platforms for supporting the advancement of local Indigenous communities, especially youth. In all cases, a major motivation for establishing (The Bishop and 'Imiloa) or transforming (Te Papa) the science centre/museum was a response to the effects of colonization on the culture and knowledge of each land's Indigenous peoples. It is significant to note that all three of these museums are located where there has been (and continues to be) strong national movements to protect, preserve, and renew Indigenous knowledge and cultures. Certainly, these larger social movements contributed to the development of these institutions. A critical point of reflection is whether there would be *any* examples of science centres in this mould were it not for successful resistance movements by Indigenous peoples.

Potential futures for inclusive science centre practice

In this chapter, we have explored how science centres have historically been, and continue to be, platforms of cultural imperialism, participating in advancing Western Science as the only legitimate worldview in the modern human narrative.[14] This paradigm is no longer acceptable, and it alienates visitors from cultural and ethnic backgrounds that have been oppressed through Western colonization. Even the seemingly innocuous mission statements – "to inspire the general public to create positive relationships with science" – serve to deny the historical reasons why certain communities are alienated from Western Science narratives.

There have been, and continue to be, efforts by science centres to address these challenges. The exemplars are few, relatively new, and varying in format, impact, sustainability, and visibility. Despite these variations, there emerge some guiding principles to inform efforts to decolonize science centre practices, especially with respect to Indigenous Knowledge Systems:

- There is a need for investment of both human and financial resources to build the capacity for understanding multiple scientific worldviews and to build authentic, collaborative relationships to enable co-development of content.
- We need to prioritize adhering to cultural protocols of Indigenous Knowledge Systems, especially with respect to considerations of intellectual property rights.
- We need to clearly differentiate between strategies that serve the science centre's goals and those that serve the Indigenous communities' goals.

Programs that directly serve Native youth, professional development interventions, and exhibits and programs that provide equal integration of Indigenous Knowledge Systems and Western Science have not yet disrupted the mainstream narratives of science centres. But they do reflect the flexibility and creativity to create multiple entry points inherent in science centre practice. In this way, they begin to approach the issue and broaden engagement with science to fully embrace and celebrate the diversity and richness of human scientific thought and innovation.

Notes

1 United Nations Declaration on the Rights of Indigenous Peoples. 2008. Official Records of the General Assembly 53(A/61/53), part one, chap. II, sect. A, available at www.un.org/esa/socdev/unpfii/documents/DRIPS_en.pdf.
2 Dawson, Emily (2018). Reimagining publics and (non) participation: Exploring exclusion from science communication through the experiences of low-income, minority ethnic groups. *Public Understanding of Science*, 27(7): 772–786.
3 Meluch, Wendy (2014). *Bishop Museum NHEP Project On Site Fieldtrip Summative Evaluation Report*. Visitor Studies Services, available at http://informalscience.org/sites/default/files/BishopMuseum_DoE_NHEP_FieldTrip_SummativeEval_2014_0.pdf. Bouman, Katherine S. (2006). Past and present tense: Understanding the visitor experience in the indigenous Australians exhibition at the Australian museum. *Visitor Studies Today*, 9(2): 11–19.
4 Dawson, Emily (2018). Reimagining publics and (non) participation: Exploring exclusion from science communication through the experiences of low-income, minority ethnic groups. *Public Understanding of Science*, 27(7): 772–786. Rennie, Leonie J., and Williams, Gina F. (2002). Science centers and scientific literacy: Promoting a relationship with science. *Science Education*, 86(5): 706–726.
5 Maas, Ad (2017). Introduction: History of science museums between academics and audiences. *Isis*, 108(2): 360–365. Wingert, Kerry (2014). Designing programs that value traditional ecological knowledge. *Relating Research to Practice*, available at http://rr2p.org/article/357

6 Salmon Camp, available at http://informalscience.org/salmon-camp-research-team-renewal-2009-evaluation-report Bishop Museum Field Trip Report: http://informalscience.org/sites/default/files/BishopMuseum_DoE_NHEP_FieldTrip_SummativeEval_2014_0.pdf

7 Maryboy, Nancy C., Begay, David, and Peticolas, Laura (2012). *The Cosmic Serpent: Bridging Native Ways of Knowing and Western Science in Museum Settings*, available at http://cosmicserpent.org/uploads/downloadables/CS-Leg acyDoc27Nov2012.pdf.

8 Native Universe, available at http://nativeuniverse.org/.

9 Wingert, Kerry (2014). Designing programs that value traditional ecological knowledge. *Relating Research to Practice*, available at http://rr2p.org/article/357. Bouman, Katherine S. (2006). Past and present tense: Understanding the visitor experience in the indigenous Australians exhibition at the Australian museum. *Visitor Studies Today*, 9(2): 11–19. Weinberg, J. (2004). People to people: Tying science to culture in South Africa. *Dimensions* (May–June): 8–9. Kelly, Lynda (1997). Indigenous issues in evaluation and visitor research. *Visitor Behavior*, 10: 24–25.

10 Oregon Museums of Science and Industry. (2016). *Generations of Knowledge: Traditional Ecological Knowledge and Environmental Science*. Portland: Oregon Museum of Science and Industry.

11 Yupik Exhibit: Overview, available at http://informalscience.org/yupik-science-and-survival-old-tools-new-knowledge-planning-grant. http://informalscience.org/yuungnaqpiallerput-way-we-genuinely-live-masterworks-yupik-science-and-survival-0

12 Maryboy, Nancy C., Begay, David, and Peticolas, Laura (2012). *The Cosmic Serpent: Bridging Native Ways of Knowing and Western Science in Museum Settings*, available at http://cosmicserpent.org/uploads/downloadables/CS-Leg acyDoc27Nov2012.pdf

13 Meluch, Wendy (2014). *Bishop Museum NHEP Project On-site Fieldtrip Summative Evaluation Report*. Visitor Studies Services, available at http://informalscience.org/sites/default/files/BishopMuseum_DoE_NHEP_FieldTrip_SummativeEval_2014_0.pdf. Huerta Migus, Laura (2011). *Shifting Paradigms: Embracing Multiple Worldviews in Science Centers. Dimensions*, (November) Washington, DC: Association of Science-Technology Centers.

14 Battiste, Marie (2016). Research ethics for chapter protecting indigenous knowledge and heritage. In *Ethical Futures in Qualitative Research: Decolonizing the Politics of Knowledge*, edited by Norman K. Denzin and Michael D Giardina, 111–132. New York: Routledge.

9 Navigating worldviews

Laura Huerta Migus in conversation
with David Begay and Nancy Maryboy

Laura *Huerta* *Migus:*	In the previous essay, I discuss the challenge of science centres' baseline Western orientation to science, and how it presents a challenge to truly reaching multicultural audiences. Both of you have personal experiences in working with and in Western science systems to "legitimize native knowledge." Can you tell me a bit about that experience?
Nancy *Maryboy:*	When we started, we were funded by the National Science Foundation (NSF) to do the first project that started this trajectory, which was the *Cosmic Serpent* project. What comes back to me is [that] we had no role model to follow at all, because almost nothing like this project [working with science centres] had been done in this field before. There was one other NSF-funded project that had to do with Native voice and knowing, led by Bonnie Sachatello-Sawyer, which was very successful. So, we were treading new ground, and there was almost nobody we could turn to for advice. We had a wonderful team, which included you, Laura, and I think we accomplished a tremendous amount, including publication of the book *Cosmic Serpent: Collaboration with Integrity*, which has served as a guide for many people doing this kind of work since. It was a real learning process for us. We called on people we had long-standing relationships with as our mentors, our friends, our presenters, and participants. Two of these people, Isabel Hawkins and Rose von Thater, had experience leading a program supported by NASA that brought together NASA scientists and Native Knowledge holders. There were many challenges in the beginning, but communication began to improve as time went on.

When the *Cosmic Serpent* project started there was a lot of distrust, even for us, in the Native communities, and we had to deal with it. Community leaders questioned us about why we were doing this project, and then, specifically, why did we name it *Cosmic Serpent*? This last question arose because in a lot of tribes in Australia and in the US, serpents are something you stay away from and respect, but you don't deal with them.

So, we got some real criticism on this, but we explained that we were using a more global perspective, that the serpent has many different definitions in many different cultures, and we wanted to leave it open to participation. When some of the tribal people came to the first sessions of the project, they came with some distrust. But, after they'd been there a couple of days and saw how we were doing it with the utmost respect, and guided by Native Ways of Knowing even more so than by Western science, they started to feel comfortable and participate more. There were some emotional reactions the first couple of sessions we did for *Cosmic Serpent*, which were week-long workshops, and we had to deal with that emotion.

We learned that it was beneficial to have a psychologist as a participant, who happened to be a board member of the Indigenous Education Institute who was trained in mediation, and we also found it was especially beneficial to include Native counsellors who were skilled in mediation and de-escalation of disharmony. So that became one of our working strategies. We had the right people in place that could deal with these things, and again, we learned the hard way. But that's something that really worked for us.

As time continued, and the same people got back together, and as you said in your chapter that by the time of the third *Cosmic Serpent* conference, you could see people from different backgrounds and regions beginning to have relationships with each other and becoming friends. In the end, we saw, for example, a museum director working with an educator from an Indigenous school in the same community, making relationships and initiating new projects together.

David Begay: My experience and my take on this was that right from the onset we were faced with two knowledge systems, two groups of people. One group was people from the

Western science world and the other were Native people – who were predominantly the cultural knowledge holders, with some also being scientists. We were also faced with the challenge that the Native American world is conceptualized by non-natives as "one world" in this paradigm when, in reality, there are many ways of knowledge in the Native American world, tied to the empirical knowledge and experiences of each culture's environment and land. When we met with Native people from the Northwest, for example, they talked about fish – the importance of access to fish, and the impact of commercialization and infrastructures on access and their ways of life. In the Arizona-Sonora desert, people talk about the desert life, the medicinal plants that were there before the coming of the Euro-American. Before the coming of Safeway and all of that, everything that was needed was in the desert. This traditional knowledge informed their lifestyle and taught how to live with the seasons, including the rain cycles and the monsoon, and places where they can also plant beans, squash, and corn. And some people knew about the land, the watershed. They knew, from experience, years of experience and observation, where the water runs. There are always resources there to rely on.

The knowledge [that] came through all those experiences – living with the land – that is empirical knowledge, in that way coming from the desert. And when you go up to the Northwest, it's the knowledge that comes from that land and that environment. So even though we say Indigenous knowledge, it's very diverse. So trying to work with the diversity of people on the Native side was challenging.

In addition to non-natives treating the Native American world as one homogenous group, Native people sometimes do the same with Western science, but Western science is also diverse as is Indigenous knowledge. We had people in *Cosmic Serpent* that were trained in very different scientific disciplines. So the scientists would say, working across disciplines, and then going from one knowledge system to another knowledge system, was challenging. There was a lot of support needed for cross-cultural communication: on one side, Native people were talking about how things are so interconnected; on the other side, it's a paradigm based on Cartesian "draconian" methods where that kind of training, it's talking about the pieces, not so much interrelationship of the whole thing. So we had to educate each other. And we were also challenged by both the scientists and the Native people, and we were asked several times, "Why are you doing this project and what do you hope to accomplish?"

From the Native side they were more concerned about the usefulness of the knowledge and the collaboration. Will it help the Native community, the applicability of this traditional knowledge in modern times, working in a science centre and trying to get a voice to a science centre, or include it in a science centre? What are you going to do with this knowledge? People considered some of this knowledge intellectual property and didn't want to just openly start talking about [it] with others. We had to be very clear about our goals and create new language on how some of this knowledge can be useful and benefit people, especially the Native people. It was quite a challenge.

For those coming from the Western science side, we're telling people we [Native people] have this knowledge system that you don't know anything about. And we were asking them to make space within their science museum for this knowledge. I guess for a scientist who never really interacted with a Native community, he or she would be clueless. They would not know where to start. The *Cosmic Serpent* project was essentially a professional development program to provide some of the fundamental values and principles on how to collaborate with integrity, and then also to provide awareness on how these collaborations had to be carefully cultivated, at their own pace, and through different processes depending on who you work with.

To ask these people to provide voice within an already established science centre was also a challenge. We relied on the people that we worked with through the *Cosmic Serpent*, hoping that they would make that initial contact, that initial education within their institutions. Many of these people said, "It sounds just great, and we could pick up a lot of information here at these *Cosmic Serpent* workshops," but when you go back, these people that they work with back at their institutions have no understanding of what Native knowledge is. And they told us [that] they felt like they were not adequately equipped to talk about it. When these workshops happen, the *Cosmic Serpent* workshops, people, because of how they are engaged, make it simple and something that feels like it's doable, something that can be accomplished. And then you go home and you have to face reality back home.

When people don't have a background – a way of life or certain language – when people don't have that experience, it's extra challenging to try to talk about Native knowledge with others. I think it's the same with the real traditional Native people. You know, science is compartmentalized and they follow a method. It's not how the language is spoken in the tribal communities. So, there's a superficial understanding and everybody is trying to be nice to each other, trying to be more clear in their language and to really understand each other. You know, there's a lot at stake. We need real communication, real dialogue, for working collaboratively with integrity. That's hard to achieve. This need became very clear during the *Cosmic Serpent* experience.

Laura: I think there is something for me that stands out from what both of you shared about *Cosmic Serpent* and some of the other work I didn't remember. Even though, on the surface, these efforts have been about bridging Western science and traditional knowledge systems, so much of the work had almost nothing to do with science content itself but was about "people" work. It was all of this human work around trust and collaboration and integrity and emotions that have nothing to do with science and Native ways of knowing. Has that been true throughout the various projects that you've been a part of working with science centres and museums?

Nancy: You know, at the beginning of *Cosmic Serpent*, we made sure we gave space to more scientists, and we probably did more of that in *Cosmic Serpent* than in the rest of our projects. The reason was that we so believed that there's a dominant society in this country that is promulgating their own sciences, through grade school, high school, college, and everyone – even Native people – are supposed to learn that. So, there was very little knowledge of Native, Indigenous knowledge systems. And to balance that we had to teach about the Indigenous knowledge content. And this was challenging because a lot of the Indigenous knowledge is proprietary to the tribe. It is not easy to scale up. It is not even recommended to share it out of context because Native people have historically been hurt by what they've shared.

Basically, we knew we had to do some teaching. An eighth of the people that came didn't even know there are still Indigenous people in the United States, so we had a huge learning curve there for people. And then, as you described in your chapter, science is mostly talked about in terms of a Western system or Western knowledge, and so when you say the content of science, one of the other things working against that was that most of the Native people were much more interested in the relationship building, and that didn't just mean with institutions. It meant among the individuals. And so there was more interest for us in seeing the Indigenous presentations rather than listen [to] someone talking about hard science. And when you take these things out of context, whether it's Native ways of knowing or hard science, you know, just a little sound bite of something, that's all it is. It's a bite. But, if you added that together with the fact that most of the Indigenous people were more interested in meeting other people of colour and other underserved minorities, and that's still true today, you know, I think you're right in saying these experiences are more than just professional development projects or systems. We're trying to make relationships.

David: You have to build a relationship. In many places there was no historical relationship. For example, the Arizona-Sonora Desert Museum: when we went to the Native people, they said, "We knew the museum was here, but we were never ever invited. We felt that we were not welcome there. And so, we drive by there from our reservation, going over to Tucson many times, so we usually drive by, a couple of miles away. We look this way. We see the museum facility, and we didn't know each other, even though we were close geographic neighbours." So, in my particular evaluation of all of this, it told me it takes two to three years just to build this relationship and to build this trust, just as a first step. Then, you know, once you've built the trust and you agree to work together, *then* you start talking about the content. What would be allowable to put into your science museum? What would be useful or would be appropriate? Then you start talking about that, and for what reason? You want to benefit from this working relationship. You don't want to just give, give, give, give.

And what I have heard at some of the *Cosmic Serpent* gatherings from Native people is, "we're talking about our

[Indigenous] knowledge more than the science is talking about how they can maybe work together. They're more reserved, and it seems like we give them more content. If we're having dialogue, then let's do it on an equal basis." Eventually, I assume there can be a point where you can get a little bit more integrated in your open relationship and build really solid knowledge that has a Native voice, and in the Native voice there is integrity and knowledge of the Native way of thinking, and at the same time it doesn't diminish the Western knowledge and expectation and efforts. So they complement one another. It's just a different way of looking at the same thing. It's just a different approach.

Nancy: But something else I want to just put in here that's really important, I think, is that how much of this was built on relationships. In every instance, at every science centre and at almost every tribal museum, we had allies, and there might be two or three. We used to call them "bridge people," and now we think it's just much bigger than just the museum world. This is the whole world, but these are allies that really care and really have knowledge, and carry parts of both ways of knowing, and it's part of their life. You know, Laura, you're one of them. Isabel Hawkins is one. Laura Peticolas is one. We have them all over the place. I can't tell you how important that is to this kind of work.

Laura: Thank you for sharing that. One of the things that stands out, as you were talking about the progression of the work, is the number of years of relationship building required before we got to content. And I think that even in the dialogue today, in working with all different kinds of museums, we always want to say that partnership takes time. But I do think your work is a good example of just how much time it probably should take you if you're starting from scratch. And that was with a lot of resources! I think it's easy to think 'We'll just partner!' But it can feel very transactional and hard to maintain partnerships.

Nancy: You mentioned some of the ways that it's difficult to carry this on. Time is one, because science centres have a Western schedule they are adhering to, and the Native people have a totally different one. The Western people have to be responsible to their board, but the Native people have to be responsive to a whole tribe and community, and all different hierarchies, different departments, etc.

And then there's the issue of money. When a project ends, in some science centres, most of the people over there are working in Western kind of silo systems. So, when there's no money to be put to something, the work comes to an end. On the other hand, some museums are so committed that they keep up things that they've just taken over and funded internally.

I think my big piece of advice would be that if you're going to work with Indigenous people you need to truly co-create this from the start. And that doesn't mean you going to them and saying, "I've got a project. There's this money out there. We can do it if we do it together." That's not what I mean. It's going to an ally in a tribe and saying, "What are some big needs of yours and how can we help fulfil them in this world of science centres and tribal museums?"

All the way along we knew what the Western museums would get out of this, but it was very hard for us to even pinpoint what the Native museums would get out of it that would be of real value to them. On the Western side of things, the assumption is that knowledge is free to give away. So incorporating new knowledge is a benefit. But on the Native side the assumption is different. Knowledge is closely held and guarded. So how do you come to terms with the two different ways of protecting knowledge when one is based on an individual getting a patent and the other is based on a whole tribe holding knowledge, with their own negotiations about access to this knowledge?

There's a very good organization called ATALM, Association of Tribal Archives, Libraries, and Museums, and they have 250 members, all operating from a very Indigenous paradigm. They've reached out to David and I because they wanted to incorporate some more STEM into their own museums and libraries. This was, to us, like a sea change in attitude in Native communities, an openness to incorporating things from each way of knowing, that they could reach out and help each other.

The final thing I wanted to say is that there is a lot of interest today in diversity. That's probably a better word to use than bridging Indigenous ways of knowing and Western science. The whole conversation around diversity, which is a huge topic, is really relevant to what we're talking about. It seems that once you open it up and call it diversity and access for different minorities into local museums, big or small, then you're kind of bumping up the conversation to another level, and it's a level that more people can participate in.

David: I think whenever you engage with Native people, there's going to be some degree of acculturation and assimilation into the dominant way of thinking, the world view of Western science. Today, when you see a Native person, that person is usually trained in Western science. You know, K-12 schools, they are very similar in all the states of the United States. And when you go to college, it's still very similar, based in Western science. And so thinking that this Native person is going to be walking around with a lot of traditional knowledge sometimes is not true. You have to contact the right people, those who hold whatever knowledge that you're going to work on. I think it helps to pull together an advisory group of Native people, not just pick somebody because they're Native, but Native people that are knowledgeable in certain areas, so you'll be given advisement from the right combination of people.

Laura: In this book, we are really trying to shake something up for practice in science museums. Not just a critique of blind spots, but a friendly critique, like when a family member tells you, "Okay. Well, I know you don't want to know this about yourself but you need to, to be healthy and survive." In that spirit, are there things that you feel are important for people in science centres and science museums to hear in that spirit of wanting them to be healthy professionals as well as healthy organizations?

David: I think one of the knowledge and traditional knowledges that may be useful in these scientific museums would be ecological knowledge. You know, the native people always bring to the table that things do not happen in pieces in the natural world. The natural world is a world that's interconnected, like a web. It's so complex that any one thing will affect the other. Trying to think about this through a cause and effect, one thing at a time frame, that's not how nature works. One little thing you do is going to affect everything, the evolution of a whole system. What Native people always bring to the table is this notion that all things are interrelated, as well as the human connection.

You know, we're now living in a world that's toxic. We pollute our water. We pollute our air. We pollute just about everything. We've polluted the plants and then the plants that the animals rely on, we polluted them, so the animals' bodies are also polluted. And then we also consume those animals. And then the manifestation of this whole process is our body is weakened.

So it comes home, and all of a sudden we're raising that question, "What happened here?" Soon we're living in a chemically saturated world. We're living in a polluted world. And if your water is sick, you're going to be sick. If your land is sick, you're going to be sick. If your air is sick, you're going to be sick. These things don't just happen out there. They're so interconnected. The air we breathe is the air out there. The water that we drink is not "over there" but is directly connected to the human body. We're so closely connected.

These kinds of teachings may be useful because we live in one world. We've got only this planet and right now it's really on a course that may not sustain us. And where do you go for this knowledge, this kind of thinking? Native people, we have a lot of that knowledge. And so, using traditional knowledge and putting it back into science, so people can become aware that we have to be more careful and we have to live in harmony with our environment. And there's nothing mythological about that knowledge. There's nothing mysterious about it. It's real.

Nancy: I am totally convinced that this way of seeing everything as interconnected kind of process is the only thing that's probably going to help, and we're going to have to jump out of this little box really fast, because this whole denigration of our planet is happening. Federal and some state policy priorities are changing and not necessarily for the good of the planet and the well-being of the people. Many decisions that take place at some federal departments are causing ecological damage, harming where we are and speeding up this process of increasing toxic waters and toxic air and everything.

I happen to firmly believe that this is really necessary knowledge, so listen to what the Native people are saying. And the discussion goes down into languages of quantum physics and systems theory and complexity theory and chaos theory. Maybe we're living more in the chaos theory actually, right this minute. But I think it's more important than ever that these knowledges be shared, and to look at what different tribes are doing to restore their waters and their plants and their air and take a lesson from that. And maybe it's time for people to let go of some of their long-held assumptions about science.

10 Youth and the challenge to define the museum of the future

Hooley McLaughlin

I remember the moment when I realized that everything I had done as a museum curator might have sprung from an egocentric perspective. It was when I was visiting Newtown, consulting for what was to become the SciBono Discovery Centre a few years later. SciBono is a successful interactive science museum built into an old turbine building, the Electric Workshop, but at that time, in late 1997, many of the large industrial buildings in the area were derelict and the area was filled with makeshift shanty homes. Indigenous African families lived in extreme poverty in this central-Johannesburg district. Violent crime was at an all-time high, and yet despite the dire social conditions there were well-funded cultural programmes underway for the youth in the area. Behind the abandoned turbine building in a series of smaller outbuildings I met innovative teachers, community workers, and, most importantly, active young people.

One programme caught my attention. A team of teenagers was producing films using donated equipment. They were taught the basic use of the cameras, lights, and sound equipment, but the content decisions were made by the students collectively. I perceived a stark contrast between this politically savvy youth group, learning the basic science and technology of filmmaking in order to explore, and effect changes in, their local environment, and our team of museum experts making plans for a standard interactive science museum. It was the young people's active disinterest in a science centre being planned for them by experts like myself that really struck home for me.

I had the credentials, I thought at that time, to be considered an expert who should be called upon to create museum displays and programmes for people in societies such as the one we found in central Johannesburg. One year before my South African visit, I had been proud to attend the opening of *A Question of Truth* (*AQOT*) at the Ontario Science Centre in Toronto. It had taken over five years of my life as a curator and the devotion of many researchers, designers, writers, interpretive planners and fabricators. The exhibition examined the bias that underlies Western science practice and application, showing how cultural perspectives dictate the questions and the tools used for scientific discovery and the accumulation of knowledge. AQOT was not only a celebration of diversity in scientific thinking, however, but also a harsh critique of the misapplication of Western beliefs. We did not reveal this more difficult message right away to the visitor. We felt it was important first to give examples of the diverse origins of our collective scientific knowledge. We showed how South Sea Island navigation had been a sophisticated system of philosophy and applied science that had resulted in large-scale ocean travel for the past millennium, a feat that would not be equalled by Europeans until the mid-1700s. We demonstrated how the use of human psychology in medical treatment was practiced at an advanced level in African societies for generations, well before the concept occurred to Western medical practitioners. We displayed evidence that European discoveries in physics, chemistry, medicine, mechanics, astronomy, and optics were predated by forgotten advances in Islamic culture. We presented displays that explored medical practices and herbal remedies that were developed and used successfully in East and South Asia long before similar discoveries and approaches were investigated in Europe. We displayed early advances in astronomical observations that were used to create precise calendric predictions of solar, lunar and planetary motion in the Middle East, South Asia, Africa, Northern Europe and the Americas, dating back to between 500 and 2500 years ago. We explored the bias revealed by our unconscious male-centric viewpoint when we consider sexuality, conception, birth, and parenthood. We presented examples of how the Indigenous cultures in the Americas, Oceania, and Africa had developed advanced knowledge of ecology, mathematics, engineering, metallurgy, navigation, astronomy, medicine, agriculture, and sociology as well as highly effective civic governance, well before European colonization.

Notwithstanding those salutary reminders, however, the tougher message was that Western science has had a dark side that we are often afraid to confront. AQOT was structured around bringing to light a prejudice that underlies all Western society: namely that Western science carries with it the conceit that there is no more valid method that can be used to decipher Nature. The outcome of this attitude is the misguided belief that the development of Western science is proof that Europeans have a superior civilization and are, in fact, themselves the highest manifestation of *Homo sapiens*. The exhibits were designed to be experienced in a somewhat linear fashion, thus bringing the visitor this message in incremental doses. The closing exhibits hit hard, though, describing how a belief in European, white, male superiority resulted in the use of science as a tool for prejudice, discrimination, and atrocities, such as slavery, Nazi death camps, genocide of Indigenous peoples around the world, and an almost universal subjugation of women. We invited visitors to participate with their comments, and felt the reactions indicated that we had made our point.[1]

Significantly, within the first year following the 9/11 attack, a travelling version of AQOT was staged at the Liberty Science Center, Jersey City, with a view of the former World Trade Center, and in 2004–2005 was presented for over a year at SciBono in Johannesburg. Despite the emotionally charged, and largely positive, reactions of visitors, however, in academic, journalistic, and museum-professional circles the exhibition was criticized for being anti-scientific, too culturally relativistic, and "politically correct." And interestingly, there were critics who felt that it was not relativistic enough, and that it subtly and perhaps somewhat secretly, despite its pretentions, supported the long-held bias that Western science is still the criterion against which all knowledge must be measured.[2] More than twenty years after its opening, AQOT still has a permanent version on display at the Ontario Science Centre.

For me, the curator, AQOT had had a painful birth. I spent the five years of its development struggling to divest myself of restraints that defined the beliefs and attitudes trained into me as a Western scientist. As one example, I could not initially grasp that the centre of any system is arbitrary and depends on what you are looking for and from where you are looking. Thus, a sun-centred planetary system may be useful for the study of gravity but is useless for making calendars and navigational systems. The outcome of this revelation was an exhibit that challenged the visitor to consider that the long-held belief that Galileo's defence of the Copernican system is only of limited importance, and that our orthodoxy in this instance had precluded our collective appreciation for the advanced thinking exhibited by South Sea Island navigators, Mayan calendar makers, and pre-Copernican astronomical observers in Asia, Africa, the Middle East, and Europe. In retrospect, it is easy to see why some had considered us anti-science, given that we had criticized Galileo in an exhibition that opened only four years after Pope John Paul II had officially apologized on behalf of the Roman Catholic Church to Galileo for threatening him with torture in order to force him to recant his determination that he had "proven" the Copernican theoretical model showing that the Earth moves around the Sun.[3,4]

For the exhibition AQOT, the criticism of Galileo was pivotal because it opened the door to the portrayal of Western science as only one of many systems of thought. The cultural bias that lies within every scientific methodology, although familiar to students of the philosophy and history of science (see Nader[3] and Kuhn[4]), is not a subject that lends itself easily to public exhibitions. In order to reach an audience that is educated to "believe" in Western science, and in part because I was myself also struggling with cultural relativism, I chose to be somewhat conservative in our exhibit design. I did not question a hard reality that exists outside of the observer's mind, for example, as many philosophies move us to consider. Rather, I stuck to a fairly orthodox view of the world in order to make the exhibition approachable for the general Western public. Within this perspective is the argument that what we observe does not change when we shift from one way of looking to another. The planets hang in the sky in the same positions whether we choose to see them spinning around the Sun or around the Earth. Their Earth-centred spin can still accommodate their spins around the Sun too; it just looks like there are extra planetary loops as we look up at the sky from the surface of Earth. James Brown in *Who Rules in Science* was gently scathing in his critique of AQOT, pointing out that the exhibition is careful not to attack Western science and falls well within the realist camp, in harmony with both the scientific orthodoxy and the beliefs generally held by educated people in Western society. It is this conservatism that can result in the hidden message that we will always come back to Western science as the measure against which we assess all other would-be "pretenders" to the throne of scientific method. And in retrospect, I think I missed this subtle failing on my part as a curator. The implications are quite serious. It leads to the question of whether I was really listening to the arguments being proffered from non-Western philosophies of thought, and whether I was opening the door to a future reconciliation, as I had thought I was, with the world that is not male, white, and of European descent.

My failure to listen was compounded by the support I received from an advisory panel of outside consultants who met frequently during the mid-formative and final stages of exhibition development. Consisting of community leaders, educators, and thinkers from many of Toronto's diverse neighbourhoods and research/teaching institutions, this group of over 20 dedicated individuals combed over every aspect of the exhibition and debated every design approach. Added to these close observers, we consulted over 400 experts in fields of knowledge, Western and non-Western, that were displayed in the exhibition before we were satisfied (perhaps in retrospect prematurely) that we had a finished product that presented science in a manner that celebrated all ways of knowing as equal partners in the pursuit of scientific understanding. We were proud to hammer home the message that ignorance and rejection of the equal merits of the world's knowledge systems were endemic and had been responsible for the misuse of Western science over centuries, resulting in repeated episodes of prejudice, discrimination, and atrocity.

Despite all our best intentions and hard work, however, we had missed an essential ingredient for success – authorship. The importance of exhibition authorship became clearer to me when I visited the youth film group in Newtown. It was the *leadership* shown by the young people involved that was so foreign to me. These young people were living with their extended families in shanty homes made of cardboard and tin, set-up in abandoned buildings and under roadways. They had no interest in the interactive science museum that we were planning, and they did not care if planners like me were trying to incorporate African history and knowledge into the preliminary concepts for the museum. Their connection to their environment was immediate and stemmed from a culture that was living, vibrant, and adapting to a changing South Africa. Most of all, they were in charge and were not following a programme designed by a teacher, a curator, or someone else's philosophy.

I went back to Toronto questioning whether I had in fact spent years of work ignorant of my overriding need to control the message of the museum, even as I campaigned for the rights of the oppressed. Freire points out that the oppressor cannot be the one who leads if the system is really going to change.[5] I unearthed the earliest "map" of the exhibition *A Question of Truth*, one scribbled in pencil and made two years before I had brought together the advisory panel. I hadn't looked at those early notes since I had made them. It was no minor shock. Approximately nine-tenths of the final exhibition were delineated in that early version. Had I gone through an extended exercise of self-doubt, and repeated reviews by our diverse advisory panel, only to end up with my original concept? Had I listened to anyone over that time period? Had I unconsciously manipulated the exhibition process to favour my vision? If we become conscious of bias, can we endeavour to avoid curatorial control that silences the voices of those who most need to be heard?

The chapters of this book describe a future for museums that will be led by the very people that hitherto have been the exhibits, and who now aspire to be the curators of their own histories and education. The discussions allude to language, cultural references, and epistemologies that do not necessarily adhere to Western scientific principles and pursuits, and where Western science is counted as only one of many among the knowledge systems that will be represented. In the future we hope for a more level playing field and for leadership to be derived from all participants.

Despite the care taken by the writers to be inclusive (indeed, they exhibit an openness that extends to an acceptance even of the purveyors of the colonizing Western system of thought) one wonders if we shouldn't establish some ground rules to protect everyone from the controlling personalities of the Western-trained curators. In his preface to Fanon's *The Wretched of the Earth*, Sartre admonishes Europeans – you are to come close to listen to the non-Europeans talking around the fire making their plans for their own future, for the territories and properties and representations previously governed by you, the oppressor. Listen, Sartre tells Europeans, but know that they are not talking to us, about us, for us, or with us.[6] That time has passed. So, I suggest all of us read and listen and accept that the words and thoughts are not meant to fit neatly into a Western science and a Western museum. And that is fine because the leadership will be coming from the people who are now talking. The future museum will be different. The people we hear from in these chapters are reaching out to everyone in the spirit of generosity and mutual respect, but time is at a premium. The next generation will not likely be so patient. Museums are at the forefront of cultural expression and need to move swiftly to accommodate the societal changes that are soon to overwhelm us all.

Huerta Migus describes *Native Youth in Science – Preserving Our Homelands*, a collaboration with the Mashpee Wampanoag Tribe, the Woods Hole Coastal & Marine Science Center, and the US Geological Survey. Bringing together Indigenous Knowledge Systems and Western Science was successful in this case because the true leadership fell to the tribal leaders who nurtured the learnings of the youth participants. It was deemed important for the Western-trained scientists not to arrogate to themselves the power to define what is and is not correct knowledge. Perhaps even more limpid with regard to our new awakening sensibilities into cross-cultural sharing is the discussion Huerta Migus has with Maryboy and Begay about the *Cosmic Serpent* project. In this endeavour, tribal communities and science centres learn together about the links between Indigenous Knowledge Systems and Western science, while simultaneously being confronted by an occasional disquieting inter-cultural communication impasse. Beliefs and practices are not always translatable and may, at times, be ultimately incompatible. In this regard, we hear that the Western practice of assuming all scientific knowledge is *acultural* and *non-proprietary* cannot be considered appropriate when knowledge is an essential part of the social construct of tribal dynamics and inter-personal respect. To override the traditional knowledge practice of teaching, learning, and sharing would be a continuation of the colonializing, forced assimilation that has brought us to our current dysfunctional inter-cultural situation. Ongoing dialogue at least opens the door to developing respect on a more equal footing, but it will be a long process.

In calling for our espousal of *failure* as an inevitable part of our cross-cultural exchange, Shannon underlines the importance of dialogue. "Shared authority, trust, and an openness to other ways of knowing" are "essential ingredients." Acknowledging failure in the process of sharing shows an awareness that current Western museum methods cannot simply be assumed to be correct. Hitherto "best practices" in the display of anthropological studies are replaced by what Shannon is calling "tentative anthropology." But this acceptance of failure is not fearful, it is brave and it is important if we are to move together to find the definition of the museum of the future. Courage is the most salient human attribute we feel when hearing from Benson, Young Wolf and Baker "Price" as they discuss the philosophy behind their Mandan Hidatsa Arikara Interpretive Center being built in North Dakota. Courage mixed with compassion for everyone who is trying to bridge the cultural gap and attempt to restore dignity lost during the cruelty of the colonialist centuries.

The Penn Museum confronts white colonization with an unequivocal challenge to all of us to stop hesitating when considering the impact of intellectual hegemony. Scott's interview with Tukufu Zuberi, the curator of the Penn Museum's new Africa Galleries, shows that we can move to that next level where leadership is taken up by the previously oppressed. At the Penn Museum, Tukufu Zuberi is talking directly to the establishment, the white-dominated society, from within one of its major institutions.

In North America, Europe, and most of the nations where science museum traditions have flourished, there is an deep-seated ignorance of what the history of ideas owes to Islamic research and practice, especially between the eighth and fourteenth centuries, in mathematics, medicine, biology, evolution, ecology, mechanics, engineering, robotics, optics, chemistry, physics, and astronomy. In this regard, Dajani points out that Western prejudice continues, as we mistakenly dismiss modern Islamic culture for being fundamentally non-scientific. Dajani disproves this myth directly through her teaching of Darwinian Evolution in the Hashemite University in Az-Zarqa, Jordan, pointing out how terms like creationism do not translate across cultures and thereby demonstrating just how far we are from bridging the cultural gaps that separate us. It is interesting to see the parallels between the Indigenous relational knowledge and the awareness of the cultural and religious roots of science that are well-understood in Islam.

The chapters by Achiam and Rosenfeld & Blonder provide evidence that the revolution is happening even among the direct descendants of the society that spawned Western colonialism – from the youth of Western society itself. Achiam sums up the essential Western museum failure with reference to Haraway's "gaze from nowhere" critique[7] of the white male default that has gone unexamined and uncriticized for so long. She suggests that there are promising experiments that challenge this dubious tradition. In this context, Achiam introduces Rosenfeld's & Blonder's description of the Irresistible Project's invitation to students to take curatorial control of exhibitions in European and Israeli museums about sensitive subjects in science and technology. Taking responsibility for the socio-scientific impacts of science and making exhibits with the lives of all stakeholders in mind, as exhibited by the students' work, speaks to a future where we forge new levels of scientific interpretation and public education built on cross-cultural, and intra-cultural, awareness and respect.

The science education work being pioneered today, at the Mandan Hidatsa Arikara Interpretive Center, the *Cosmic Serpent* project, the African gallery at the Penn Museum, and the Hashemite University, are speaking to, and setting the stage for, the next generation of cultural leaders. As exemplified in the Irresistible Project, it is the youth of all our societies that will take the next steps toward responsible museum presentations and toward the very definition of these cultural heritage institutions. And as they have all through history, the youth of tomorrow will use everything they have, all their individual and collective energy, physical and mental capacity, to take the control that we reluctantly grant them. Thomas Jefferson, theorist on the subjects of rebellion and revolution, noted that to be successful a society depends on a change of leadership every 19 or 20 years, allowing the next generation "a right to choose for itself the form of government it believes most promotive of its own happiness,"[8] presumably within a new, changing set of values that has left the older generation in its dust.

Depending on cultural context and values, the change of leadership will take place with a rough transition or a more subtle one, but it will happen. It is our duty as museum professionals to acknowledge that such a change with respect to the relationship with the dominating society of white men is long overdue. In these chapters we see the willingness for dialogue and sharing, but it won't always be gentle. We can cite extremes where museums have been the stage for protests where young people have taken considerable intellectual and physical risks to demand change, often weaponizing their very youth and beauty. *Voina*, the activist anti-government group in Russia, occupied the Timiryazov Biology Museum in Moscow in 2008 with a sexual orgy to protest the insulting political slogans of Dmitry Medvedev. Medvedev, the protégé of Vladimir Putin, was calling for more female fertility to increase the Russian population. The implication that women were ignoring their biological place as second-class citizens was yet another obvious example of the tone-deaf and repressive attitudes of the regime already underway in Russia at that time. Among the *Voina* protest members that day were 18-year-old Nadezhda Tolokonnikova and her husband, 21-year-old Pyotr Versolov.[9] Tolokonnikova went on to be a co-founder of *Pussy Riot* and continues to fight for the rights of the undervalued in world societies. It is not surprising that *Voina* and *Pussy Riot* have chosen venues such as churches and museums as locations for protests, given that our traditional institutions now constitute the front lines of the culture wars. White male science hegemony pervades all museums in subtle and not so subtle ways. Similar to *Voina's* action at the Timiryazov, the 30-year-old Luxembourg performance artist, Deborah de Robertis, staged a feminist protest at the Musée d'Orsay in Paris in May 2014, by re-enacting Gustave Courbet's *L'Origine du Monde* in front of that famous painting.[10] And across North America, young people are tearing down historical statues in an attempt to shame the greater society into acknowledging our collective deference to past leaders who espoused racist and sexist agendas.

What can we derive from these bold moves? While it would be easy to dismiss *Voina* and de Robertis as sensationalist and therefore not to be taken seriously by cultural professionals, the very consequences of the actions, evidenced by Tolokonnikova's later incarceration in a Siberian prison and the repeated arrests and court appearances of de Robertis, indicate that these young artists are anything but frivolous. In the chapters of this book, we present discussions and experiments where polite and somewhat deferential steps are being taken toward the societal changes that the young artists in Moscow and Paris are shoving in our face. Perhaps the thoughtful, less sensational approach will have sustainability and will prevail over the long run, but the peaceful dialogue and the careful efforts toward sharing will be for nought if we do not realize how serious this call is for a radical shift in our collective attitude. Ultimately, it is not about sharing from a safe position where the perspective of our traditional museum methodology is maintained unchallenged and unchanged. Rather, it is about turning over the leadership to the next generation, and especially to the youth of both sexes who live in the many world cultures that have been colonized and exploited by the Western white male science hegemony. If we do not heed the call for change, we can predict that the culture battles in the museums will become increasingly confrontational.

The projects outlined in the chapters of this book describe the efforts of caring people from many cultures who are paving the way for the young people of the next generation. The future society will be different. Museums can only be a vital part of that society if we allow the mantle of leadership to fall to the people who are redefining the collective, increasingly overlapping, cultural mosaic of the world. As Judy Diamond points out in the opening chapter, inclusion demands "doing with," not "doing for," and we hope that the new leaders of the museums of the future will take up the challenge and afford themselves of the opportunities being pioneered today by a new, broadly inclusive, contingent of diverse museum professionals.

Decolonizing the museum is difficult. At times it is necessarily impolite, it must be. The diverse leaders of the projects featured here along with many others are going into territory that we never thought possible just a few years ago. And, in part because of these efforts, I believe that we can rely on the youth of tomorrow to challenge the authorship of exhibitions and the definition of the future museum.

Notes

1 Livingstone, Phaedra, Pedretti, Erminia, and Soren, Barbara (2003). Visitor views of science brought to light in the exhibition a question of truth. In *Le Musée à la Rencontre de ses Visiteurs/ The Museum Reaching Out to Its Visitors*, edited by Anik Landry, 69–76. Livingstone, Phaedra, Pedretti, Erminia,

and Soren, Barbara (2001). Visitor comments and the sociocultural context of science: Public perceptions and the exhibition "A Question of Truth". *Museum Management and Curatorship*, 19(4): 355–369. Livingstone, Phaedra (2007). A question of truth: "Writing into an Exhibition". In *Visitor Voices in Museum Exhibition*, edited by K. McLean and W. Pollock. Washington, DC: Association of Science and Technology Centers. Pedretti, Erminia and Soren, Barbara (2007). A question of truth: A cacophony of visitor voices. In *Visitor Voices in Museum Exhibition*, edited by Kathy McLean and Wendy Pollock. Washington, DC: Association of Science and Technology Centers.

2   Brown, James B. (2001). *Who Rules in Science: An Opinionated Guide to the Wars*, 1–3. Cambridge, MA, London: Harvard University Press.

3   The Church apologized to Galileo in 1992, in my opinion, because popular opinion had been in Galileo's favour in recent years, despite his having exhibited poor scientific logic in his arguments with the Church, which scholars noted at the time. (see Owen Gingerich, http://joelvelasco.net/teaching/3330/gingerich04-truth-in-science.pdf "Truth in Science: Proof and Persuasion and the Galileo Affair").

4   The pathway to clear thinking isn't all that easy even today. As recently as January 2008, Pope Benedict XVI was prevented from speaking at Sapienza University in Rome by students and professors protesting that he was anti-scientific. As Cardinal Joseph Ratzinger in 1990, he had written a paper criticizing Galileo's logic (Cardinal Joseph Ratzinger, "The Crisis of Faith in Science," March 15, 1990, Parma. Extracts taken from A Turning Point for Europe? The Church and Modernity in the Europe of Upheavals, Paoline Editions, 1992, pp. 76-79. English translation by NCR. – see http://ncronline.org/node/11541.), as have many other scholars over the years. Curiously, here was an example of a Pope who had more awareness of the need for clear scientific logic than the ardent students and professors of science in a modern university.

5   Freire, Paulo (1996). *Pedagogy of the Oppressed*. Trans. M.B. Ramos. St. Ives: Penguin.

6   Sartre, Jean-Paul (2004). Preface to Franz Fanon's *The Wretched of the Earth*. Trans. Richard Philcox, xlviii. New York: Grove Press.

7   Haraway, Donna (1988). Situated knowledges: The science question in feminism and the privilege of partial perspective. *Feminist Studies*, 14(3): 575–599.

8   www.let.rug.nl/usa/presidents/thomas-jefferson/letters-of-thomas-jefferson/jefl246.php. Cited on February 18, 2019. Thomas Jefferson letter to Samuel Kercheval, July 12, 1816, Monticello.

9   www.theguardian.com/world/2012/aug/08/pussy-riot-profile-nadezhda-tolokonnikova. Cited on February 18, 2019. Pussy Riot Profile: Nadezhda Tolokonnikova. The Guardian, Wednesday, August 8, 2012.

10   https://news.artnet.com/exhibitions/artist-enacts-origin-of-the-world-at-musee-dorsay-and-yes-that-means-what-you-think-35011. Cited on February 12, 2019. Artist Enacts 'Origin of the World' at Musée d'Orsay – And, Yes, That Means What You Think. Artnet News, June 5, 2014.

Index

Note: Page numbers in **bold** indicate tables.